Fort Collins Science Center

User Manual for Blossom Statistical Software

By Brian S. Cade and Jon D. Richards

Open-File Report XXXX-XXXX

U. S. Department of the Interior
U. S. Geological Survey

U. S. Department of the Interior
Gale A. Norton, Secretary

U. S. Geological Survey
Charles G. Groat Director

U. S. Geological Survey, Reston, Virginia 2005
Revised and reprinted: 2005

For product and ordering information:
World Wide Web: http://www.usgs.gov/pubprod
Telephone: 1-888-ASK-USGS

For more information on the USGS - the Federal source for science about the Earth,
its natural and living resources, natural hazards, and the environment:
World Wide Web: http://www.usgs.gov
Telephone: 1-888-ASK-USGS

The Blossom user manual and software is available for download from the web page of the Fort Collins Science
Center, http://www.fort.usgs.gov/products/software/blossom/blossom.asp

Contents

User Manual For Blossom Statistical Software

By Brian S. Cade and Jon D. Richards, U. S. Geological Survey

Introduction

Blossom is an interactive program for making statistical comparisons with distance-function based permutation tests developed by P. W. Mielke, Jr. and colleagues at Colorado State University (Mielke and Berry 2001) and for testing parameters estimated in linear models with permutation procedures developed by B. S. Cade and colleagues at the Fort Collins Science Center, U. S. Geological Survey (known as the Midcontinent Ecological Science Center prior to 2002). This manual is intended to update and replace earlier versions by B. S. Cade and J. D. Richards dated 2000 and 1999 and by W. B. Slauson, B. S. Cade, and J. D. Richards dated 1991 and 1994. We have expanded on material in earlier versions and provide documentation on new routines added since 2000. Routines added since 2000 are: double permutation (HYP/DP) procedures for linear model tests (OLS, LAD regression, and quantile rank score tests) when null models are either implicitly or explicitly constrained through the origin, i.e., no intercept models (Cade 2003, Cade et al. 2005, Cade et al. In press); dropping all but a single zero residual in LAD (and quantile) regression permutation tests of subsets of variables in multiple regression models (Cade 2005, Cade and Richards In press); and computation of all quantile regression estimates (LAD/ QUANT = ALL). In addition, we now offer the option of saving output to a terse formatted file that is useful for summarizing results of multiple simulations (OUTPUT /TERSE or VERBOSE), and the option to store (SAVETEST = *file name*) the vector of permuted test statistic values from Monte Carlo resampling approximations of probabilities. The computer code has been made more efficient where possible and was compiled with Lahey Fortran 95 to dynamically allocate memory.

Routines added between 1994 and 2000 included a permutation version of ordinary least squares (OLS) regression that parallels the least absolute deviation (LAD) regression permutation test; a permutation and asymptotic chi-square approximation of P-values for a rank score statistic for regression quantiles; empirical coverage tests for univariate goodness-of-fit and g-sample comparisons that are extensions of the Kolmogorov-Smirnov family of statistics for comparing cumulative distribution functions, including an option for testing goodness-of-fit for a random uniform distribution on a circle; a second option for standardizing multiple dependent variables in multiresponse permutation procedures (MRPP) based on the variance/covariance matrix (Hotelling's commensuration); computing exact probabilities by complete enumeration of all possible combinations for small block and treatment designs in multiresponse randomized block permutation procedures (MRBP); a Monte Carlo resampling approximation alternative for all the MRPP family of statistics (MRPP/NPERM); and multivariate medians and distance quantiles (MEDQ) to be used as descriptive statistics with MRPP analyses.

The permutation procedures in Blossom can be used for comparing data obtained in familiar survey sampling and comparative experimental designs.

1. Multiresponse permutation procedures (MRPP) are used for univariate and multivariate analyses of grouped data in a completely randomized one-way design. MRPP are used for comparing equality of treatment groups analogous to one-way analysis of variance (or t-test) for univariate data, or multivariate analysis of variance (or Hotelling's T^2) for multivariate data. The default Euclidean distance function in MRPP provides an omnibus test of distributional equivalence among groups or a test for common medians if the assumption of equal dispersions is applicable. Options allow MRPP to perform permutation (randomization) versions of t-tests, one-way analysis of variance, Kruskal-Wallis tests (for ranked data), Mann-Whitney Wilcoxon tests (for ranked data), and one-way multivariate analysis of variance. Options in MRPP also allow you to truncate distances to evaluate multiple clumping of data, establish an excess group, and select arc distances to compare circular distributions of grouped data. Multivariate data are commensurated (standardized) to a common scale but an option allows you to turn off commensuration. Commensuration can be done by using average Euclidean distance (default) or the variance/covariance matrix for the dependent variables. Multivariate medians and distance quantiles (MEDQ) are provided as estimates to be used in describing distributional changes detected by MRPP analyses.

2. Multiresponse permutation procedures for randomized blocks (MRBP) are used for univariate and multivariate analyses of grouped data in a complete randomized block design. Again, the default Euclidean distance function provides an omnibus test for equivalence of distributions or common medians if the assumption of equal dispersions is satisfied. Univariate comparisons are analogous to analysis of variance or Friedman's test (for ranked data) for complete randomized block designs. Options allow MRBP to perform permutation versions of these two tests. Options also allow for aligned or unaligned data analyses and to commensurate or not commensurate multivariate data. MRBP also can be used to calculate agreement measures among blocks. A linear transform of Pearson's correlation coefficient and a permutation test of significance also can be calculated in MRBP.

3. The permutation test for matched pairs (PTMP) is a special case of MRBP, univariate data in two groups and n blocks, used for paired comparisons. Options allows PTMP to perform permutation versions of paired t-tests and Wilcoxon's signed rank test (for ranked data).

4. Multiresponse sequence procedure (MRSP) is a special case of MRPP where first-order sequential pattern of data is tested against the null hypothesis of no sequential pattern. Univariate analyses are analogous to the Durbin-Watson test for first-order serial pattern and bivariate analyses are analogous to Schoener's t^2/r^2 statistic (Solow 1989). Permutation versions of these two tests can be done. Options allow you to select the sequencing variable and to turn off multivariate commensuration.

5. Least absolute deviation (LAD) regression is an alternative to ordinary least squares (OLS) regression that has greater power for thick-tailed symmetric and asymmetric error distributions (Cade and Richards 1996). LAD regression estimates the conditional median (a conditional 0.50 quantile) of a dependent variable given the independent variable(s) by minimizing sums of absolute deviations between observed and predicted values. Options allow for testing all slope parameters (full model) equal to zero or to test subsets of parameters (partial models) equal to zero by Monte Carlo resampling of the permutation distribution (Cade and Richards 1996). LAD regression can be used anywhere OLS regression would be used but is often more desirable because it is less sensitive to outlying data points and is more efficient for skewed error distributions as well as some symmetric error distributions.

6. Regression quantiles are a natural extension of LAD regression to estimate any conditional quantile and provided as an option in LAD regression. Regression quantiles allows you to estimate any conditional quantile (say τ, $0 \le \tau \le 1$) of a dependent variable given the independent variable(s) by minimizing the asymmetrically weighted sum of absolute deviations, where the weights are τ for positive residuals and $1 - \tau$ for negative residuals. A 0.50 regression quantile is LAD regression. Regression quantiles are useful in ecological applications involving limiting factors where it is desirable to estimate functional changes along boundaries of distributions (Terrell et al. 1996, Cade et al. 1999, Cade and Guo 2000, Dunham et al. 2002, Cade et al. 2005) and for general modeling of rates of change associated with heterogeneous variation in linear models. Cade and Noon (2003) provide a primer on quantile regression for ecological applications. The LAD permutation tests of Cade and Richards (1996) have been extended to regression quantiles (Cade 2003, Cade and Richards, In press). Another permutation testing alternative also is provided that is based on the quantile rank score functions for regression quantiles (Koenker 1994, Cade et al. 1999, Koenker and Machado 1999), which maintains better Type I error rates than the Cade and Richards (1996) procedure when there are heterogeneous errors. The permutation approximation of P-values for the quantile rank score test statistic was evaluated in Cade (2003), Cade et al. 2005, and Cade et al. (In press). The P-value based on the asymptotic Chi-square approximation of Koenker (1994) is also reported and was also evaluated by Cade (2003), Cade et al. 2005, and Cade et al. (In press). Both test statistics require weighted estimates to maintain correct Type I error rates with heterogeneous distributions. It is possible to estimate all possible regression quantiles and save the estimates by quantiles to a specified file.

7. *G*-sample and goodness-of-fit tests based on empirical coverages (COV) are for univiariate comparisons of grouped data similar to the Kolmogorov-Smirnov family of statistics for comparing cumulative distribution functions (Mielke and Yao 1988, 1990). These statistics are appropriate for continuous univariate responses with no or few tied values. Options allow for testing goodness-of-fit to a uniform distribution on the unit circle, which is equivalent to a permutation version of Rao's spacing test (Rao 1976).

It is our intent that this software be considered a companion to and not a replacement of other commercial statistical software. We've consciously avoided duplicating data manipulation and graphical capabilities that are available in commercial packages such as SPSS, SYSTAT, SAS, and S-Plus. We believe that graphical exploration of data and graphical presentation of results analyzed by the procedures in Blossom are extremely important for proper interpretation of your results. The open source "R" software is especially attractive.

Appendix A lists common statistical tests encompassed by these permutation procedures. The methods contained in Blossom are presented by example. Most of the examples are from ecology, but of course the procedures in Blossom can be used on many other sorts of data.

Overview of Statistical Concepts

The statistical procedures in Blossom are distribution free in the sense that probabilities of obtaining extreme test statistic values given the truth of the null hypothesis (Type I errors) are based on permutations of the data from randomization theory and are not based on an assumed population distribution (Edgington 1987, Good 2000, Mielke and Berry 2001). In most investigations, the population distribution will never be known and assuming an inappropriate distributional model can lead to weak or invalid statistical inferences. The normal distribution is an inappropriate model for many ecological data, which often are skewed, discontinuous, and multi-modal. When sample sizes are small, large sample (asymptotic) approximations often are questionable. Permutation procedures make efficient use of small samples, because probabilities can be calculated exactly by complete enumeration of all possible combinations under the null hypothesis. Of greater importance, the permutation testing framework allows us to use test statistics based on measures of variation other than squared deviations (variances). Test statistics based on variances are derived from the distributional assumptions underlying the maximum likelihood approach. Other measures of variation may be more appropriate in a permutation test that does not require assumptions about the specific form of the error distribution.

The distance-functions that form the basis of the MRPP family of tests allow test statistics to be based on powers of Euclidean distances. The distance function between any 2 observations x_i and x_j with r response variables (dependent variables) in MRPP is defined by

$$\Delta_{i, j} = [\sum_{h = 1}^{r} (x_{h, i} - x_{h, j})^2]^{v/2}$$

where $v > 0$ (Mielke and Berry 2001). We emphasize use of test statistics based on ordinary Euclidean distances ($v = 1$), a metric measure of variation that is congruent with most data measurement scales (Mielke 1986, Biondini et al. 1988). Euclidean distances are the common geometrical interpretation of distance applied to differences between replicate data values on their measurement scale. Most conventional parametric and nonparametric methods are based on squared Euclidean distances (squared deviations are squared Euclidean distances, i.e., $v = 2$). Statistics based on squared Euclidean distances (variances) are nonmetric measures (they violate the triangle inequality of a metric) that have no simple geometrical interpretation in an r-dimensional data space, where r is the number of response variables. In contrast to Euclidean distance statistics, geometrical interpretation of variance based statistics involves distances between vectors in an n-dimensional space, where n is sample size (Box et al. 1978:197-203). An n-dimensional geometric interpretation is complex, does not coincide with the data space, and results in considerable loss of graphical information because distances between replicates vanish. It is impossible to graph individual data points in a nonmetric space to examine dissimilarities (Pielou 1984:41-46). Although we emphasize tests based on Euclidean distances, analyses based on powers other than 1 (Euclidean distance) are appropriate in some specific applications.

Euclidean distance based statistics have greater power (the probability of rejecting the null hypothesis when it is false) to detect location (central tendency) shifts among skewed distributions than do squared Euclidean distance statistics (Zimmerman et al. 1985, Biondini et al. 1988, Mielke and Berry 2001). Power to detect location shifts in symmetric distributions with Euclidean distance statistics is greater than or equal to power with squared Euclidean distance statistics, depending on distributional form (Mielke et al. 1981, Mielke and Berry 1982, Mielke 1984, Mielke and Berry 1994, 1999, 2001). Euclidean distance based statistics have better power to detect location shifts across a greater variety of distributions than squared Euclidean distance (variance) statistics. Euclidean distance based statistics also are used to detect omnibus differences in distributions, sensitive to both dispersion (variation) and shifts in central tendency (median) (Biondini et al. 1988, Mielke and Berry 1994). There is no a priori reason to presume that shifts in central tendency of data distributions characterize the only effects of interest in ecological investigations.

The permutation procedures based on distance functions are readily extended to several novel applications including, truncation of values to detect multiple clustering, comparisons of circular distributions, assignment to an excess group, agreement of values, and first-order autoregressive analyses (Mielke 1991, Mielke and Berry 2001). Each of these applications will be discussed in appropriate examples.

Medians and other quantiles are estimates obtained by minimizing sums of absolute deviations and are appropriate descriptive statistics for permutation procedures based on Euclidean distance functions (Mielke and Berry 2001). Functions are provided to estimate multivariate medians of grouped data and quantiles for distances between individual observations and their group median. This function can also be used to compute medians and any selected quantiles for univariate data distributions.

Permutation procedures for testing hypotheses in linear models are available for least absolute deviation (LAD) regression (Cade and Richards 1996), a generalization for regression quantiles (Cade et al. 1999, Cade 2003), and for ordinary least squares regression (Anderson and Legendre 1999). LAD regression estimates rates of change in conditional medians, whereas the more familiar OLS regression estimates rates of change in conditional means. Regression quantiles estimate rates of change in any selected conditional quantile (Koenker and Bassett 1978). The forms of the permutation test statistics are similar for all three of these estimation methods, and are based on a proportionate reduction in sums minimized when passing from a null, reduced parameter model to the alternative, full parameter model (Mielke and Berry 2001, Cade 2005). These tests are a drop in dispersion form. The observed test statistic, T_{obs}, equals the (sum of deviations for reduced parameter null model - sum of deviations for full parameter alternative model) / sum of deviations for full parameter alternative model; where the deviations are squared residuals if OLS regression, absolute values of residuals if LAD regression, or weighted absolute values of residuals if regression quantiles. This test statistic is equivalent to usual F-ratio used in OLS regression, except that the sums minimized are not divided by their degrees of freedom (df) because they are invariant under the permutation arguments. Hypothesis testing for all three of these regression estimates are made either by permuting the dependent variables for full model tests that all slope parameters are zero (null model includes just an intercept) or by permuting residuals from reduced parameter null model for partial model tests (subhypotheses) that some specified subset of slope parameters are zero (null model includes more than just an intercept). Extensive simulation work has demonstrated the approximate validity of permuting residuals under the reduced parameter null model when making permutation tests involving nuisance parameters in linear models (Cade and Richards 1996, Kennedy and Cade 1996, Anderson and Legendre 1999). Simulation research (Cade 2003, Cade 2005, Cade et al. 2005a, Cade et al. 2005b, Cade and Richards In press) has demonstrated that Type I error rates can be improved by using double permutation schemes when null models are constrained through the origin (no intercept) and by deleting all but a single zero residual when LAD and quantile regression null models include multiple independent variables.

All the tests described above for the linear model maintain validity of their type I error rates only if it is reasonable to assume independent and identically distributed (i.i.d.) errors. If the errors are heterogenous as happens when the variance changes as a function of the independent variables, other methods must be employed. One possibility is to estimate weighted versions of either LAD or OLS regression, where weights are selected to be inversely proportional to the square root of the variances. Permutation testing then is employed on the weighted transforms of the dependent (y) and independent (X) variables (Cade and Richards 1996, Cade 2005, Cade et al. 2005a, Cade et al. 2005b). Alternatively, for the regression quantile estimates, we provide a permutation test for the quantile rank score statistic (Koenker 1994, Koenker and Machado 1999), which is not as sensitive to heterogeneity of variances because it uses the signs of the residuals and not their magnitude. Statistical performance of the permutation test for the quantile rank score statistic was investigated by Cade (2003), Cade et al. (2005a), and Cade et al. (2005b). Weighted estimates and rank score tests were required to maintain correct Type I error rates when heterogeneity exceeded a change in 2.5 standard deviations across the domain of the independent variable. Blossom also reports the asymptotic version of the quantile rank score statistic that is distributed as a Chi-square distribution with degrees of freedom equal to the difference in number of parameters (q) between alternative (full p parameters) and null (reduced $p - q$ parameters) models (Koenker 1994, Koenker and Machado 1999).

The empirical coverage tests included in Blossom are related to the Kolmogorov-Smirnov family of tests for equality of univariate cumulative distribution functions. One-sample goodness-of-fit and *g*-sample tests exist. The coverage test statistic is based on the spacings between the order statistics. These tests provide another permutation testing alternative to MRPP for univariate continuous data. Unlike MRPP, the coverage test are not appropriate when there are many tied values, as this violates the continuity assumption. Little can be said at this time about the power of the coverage tests relative to MRPP for data for which both tests are appropriate. Go forth and investigate!

Preparing to run Blossom

Blossom runs on computers running secure 32-bit Windows operating systems, i.e., Windows XP or Windows 2000. The program is not supported, but may also run under Windows 95, Windows 98, Windows ME, and Windows NT. An installation program is provided. See Appendix B, which gives computer requirements and contains installation instructions.

Installation creates a Blossom folder with the Windows and Console versions of Blossom (BLOSSOM.EXE and CONBLOS.EXE, respectively). Access the programs from the Windows "Start | Programs | Blossom" folder. Frequent users can make shortcuts from their Windows Desktop as suggested in Appendix B. Appendix B also contains information on setting up a command prompt window for the Console version. Only one instance of Blossom can exist (only one session can run at a time).

Blossom operates on data files in the current folder (local directory). The Windows version can access data through a dialog box that allows the user to change directories. The Console version accesses local data so it should operate from within the folder where the data files exist. Both versions accept command line input and in fact, most of the function of Blossom is accessed through the Blossom command line. All general (non-statistical) commands of Blossom can be given through the command line of both the Windows and Console versions. The general Blossom commands are explained in the General Program Functions section. The Windows version allows graphical interface access to some general functions. All statistical functions must be accessed through the command line input.

The Blossom command line prompt for the Console version is a ">" character (followed by the cursor positioned to accept input). Commands in the Windows version are entered via a "Blossom Command>>" entry field at the bottom of the Blossom window. In the Windows version a copy of the command is written to the Blossom session output window.

Blossom output goes to the user console (screen), specified or implied output files, and to a session log file called BLOSSOM.LOG. Blossom writes command input and program results to the screen in the Console version and to an output window in the Windows version. Results from the statistical procedures are written to a local output file (in current folder). Blossom keeps a history of commands given during a session in the BLOSSOM.LOG file located in the installed BLOSSOM\LOG folder.

The output window of the Windows version contains both input to and output from the Blossom session. The contents of the output window can be saved, copied, or printed. The contents can be erased during a session with the CLS command explained below. Each session begins a new output window. The contents of the output window are lost when a session is quit or the CLS command is given. The screen (or console) output of the Console version also contains program input and output. The amount that can be seen or recovered during a session depends on the properties of the command prompt window (Appendix B). Access to the command prompt window is through normal Windows interface to any command session.

Results from Blossom statistical programs are written to Blossom output files. These are named and created with the USE or OUTPUT commands as explained below. The output of each statistical procedure is appended to the output file so these files may be used again as needed. The NOTE and DATE commands provide a means to annotate and write the date and time to a Blossom output file. In the Windows version, Blossom starts in the folder where data were last accessed during the previous session.

The BLOSSOM.LOG file in the installed BLOSSOM\LOG folder keeps a history of commands given during a Blossom session. This file is saved at the end of each session but is rewritten when another session begins. To retain the session history the log file contents should be copied to another folder or file after a Blossom session is quit. The quote (comment) command is a way to write documentary comments to the BLOSSOM.LOG file.

Data Formats

Blossom is devised to operate in local folders containing data files. (Output of statistical procedures is written to local Blossom output files). The Blossom user specifies the local folder either through target folders of shortcuts or the folder from which the Console version is executed. In the Windows version, folders may be changed via a dialog box to access data files (or Blossom submit files).

Blossom can read ASCII text files, SYSTAT (.SYS and .SYD) files, S-Plus 2000 data frames with only numeric values, and some Data Interchange Format (.DIF) data files. (Microsoft DIF files reverse the order of tuple (observation) and vector (variable) DIF convention. Blossom cannot read Microsoft DIF files). Only numeric values can be read and used within Blossom. Character (string) variables can't be used in ASCII text data files and character variables are ignored in SYSTAT files. S-Plus 2000 (compatibility with other versions of S-Plus is uncertain) data frames must have be identified with the name followed by "." with no extension. All numeric variables are treated as real numbers, not integer numbers. Whole numbers, however, may be entered with or without a decimal point. Numeric values including leading + and – signs, the decimal point, and the places necessary for exponential notation (if used) must not exceed 25 places. Missing values are indicated by a period (a lone decimal point). Blossom tallies the number of missing values in an analysis and appropriately remove missing cases (if possible).

ASCII Text Files

Text files can be used by Blossom if they contain ASCII text. Unicode text files can't be used by Blossom. ASCII text data files read by Blossom contain columns of numbers where each column is separated from the others by at least one space or a comma. Data are read in free format, thus columns need not be perfectly aligned. Each column in the data file contains values of a variable for each of the objects (or observations, events, or cases) sampled. Thus, there is a column for each variable and a row for each object. The variables represent different measurements or observations made on each object, and such information as to which group or block the observation belongs. Here is a sample data set to be used later.

```
1    4  5
1    3  4
1    4  3
2    2  3
2    2  2
2    3  2
2    3  1
```

It contains observations on seven objects (rows) with values for each of three variables (columns). In the data shown above the variable in the first column is a grouping variable which indicates membership in one of two groups (values of 1 or 2). The other two variables are measured values for each of the seven objects. To make this example more concrete, think of the grouping variable as indicating a burned versus an unburned site and the other two variables as the abundance of two different species. Alternatively, think of the grouping variable as indicating gender and the other two variables as measuring skull length and width.

The order of the variables (columns) in the file and the order of the cases (rows) in the file makes no difference. Blossom does all the necessary data sorting. For MRPP, where groups are compared, the grouping variable must exist in the file. For MRBP the blocking variable must also exist in the file. For LAD, OLS, and quantile regressions a minimum of two columns must be given, one for the dependent and one for the independent variable. They can be in any order.

ASCII data files can be of two different forms. The first form is merely the columns of numbers described and shown in the example above. These are called unlabeled data files. To use this form of data, names for each column (variable) must be specified in the USE command as described below. The second form also consists of columns of numbers, but the first row of the data file contains the variable names. These are called labeled data files. The labels or variable names must contain no more than 25 alphabetic or numeric characters, start with a letter, and contain no blanks or other special characters (the underscore character however is legal). In labeled data files, the labels must be separated from one another by at least one space or comma. Here is the same data from above but shown as a labeled data file.

GROUP	X_COORD	Y_COORD
1	4	5
1	3	4
1	4	3
2	2	3
2	2	2
2	3	2
2	3	1

This file is spaced so that the numbers are listed beneath the labels or variable names, but this is not necessary. The file is equally readable by Blossom in the following form.

```
GROUP   X_COORD   Y_COORD
1 4 5
1 3 4
1 4 3
2 2 3
2 2 2
2 3 2
2 3 1
```

There must be a single label for each column of data (variable), no extra labels or extra columns of numbers are allowed (there must be data for each variable). Blossom checks the data for many errors. But a data file intended, for example, to have four variables and six cases could be read by Blossom if, by mistake, only three variables were labeled and four columns of numbers existed, then Blossom would interpret the data file to have three variables and eight cases. Unless you make a mistake that evens out like this Blossom will detect the error.

Sometimes the data will have too many columns to fit on a single line. In such cases it is legal for the rows of data values and labels to be continued (or "wrap") to the next line in the data file. Even this small, sample data file can be read if one set of observations occupies more than one line in the data file. Blossom can read it in the following form.

```
GROUP   X_COORD
Y_COORD
1 4
5
1 3
4
1 4
3
2 2
3
2 2
2
2 3
2
2 3
1
```

It is possible to add comment lines to a data file by beginning each line with a single (') or double quotation mark ("). The quotation mark must occur as the first non-blank character of the comment line. Such lines are completely ignored by Blossom as command input, but are written to the Blossom.LOG session history file for documentation purposes. For example, the data given above could appear as follows:

```
' Spatial coordinates of young and old birds
' data collected summer 1989
  GROUP X_COORD Y_COORD
' begin group 1 = young
  1       4        5
  1       3        4
  1       4        3
' begin group 2 = old
  2       2        3
  2       2        2
  2       3        2
  2       3        1
```

Programs such as statistical packages, text editors, spreadsheets, database software, and word processors can all be used to produce data files in ASCII text format. (Program documentation should be consulted about how to specify ASCII text output files). Data lines should be no longer than about 4000 characters and no more than about 1000 variables. (Blossom statistical commands select subsets of variables in data file for analysis, but very large (number of columns) data files are unwieldy and should be avoided for practical considerations).

SYSTAT and DIF Files

Data Interchange Format (DIF) files with all numeric values can be read by Blossom. DIF files have the exetension .DIF, and can often be written and read read by spreadsheet and data base programs. Note that Microsoft DIF files reverse the tuples (rows) and vectors (columns) of data so they cannot be read as standard DIF files by Blossom.

SYSTAT data files (.SYS and .SYD) also can be read by Blossom. Only numeric values can be used by Blossom; character string variables are ignored if present. S-Plus 2000 data frames with only numeric variables also can be read by Blossom.

General Program Functions

Either the Console or Windows version of Blossom is invoked to begin a Blossom session.

To begin a Console session in the operating system command prompt window, type CONBLOS followed by the ENTER key. As soon as the program's prompt, the greater than symbol (>), appears, Blossom is ready to receive commands. Commands can be typed in upper or lower case.

For the Windows version, start BLOSSOM.EXE as any Windows program by clicking on the shortcut icon or from Windows Start Menu selection: "Programs | Blossom | Blossom". At the bottom of the Blossom window is the "Blossom Command>" entry field.

In this document, to make them stand out, commands are always shown in UPPER CASE red. These commands are shown with the Console version prompt, indented like this:

>USE FROG

After the command has been completely specified, enter it with the ENTER key. For most commands, only the first two or three characters need be typed, but it is good practice to spell them out completely. If a command is too long for a single line, it can be continued on the next line by entering a comma at the end of the line to be continued. In the example command lines given in this document, the > symbol should not be entered; it is supplied by the program and appears on the computer screen for the Console version. For the Windows version the input cursor must be in the "Blossom Command>" entry field.

Commands for a complete Blossom session may look like the following.

>USE FROG
>TITLE Final Analysis of North Fork Frog Study
>OUTPUT FLAST
>MRPP AGE HEALTH SIZE * LOCATION / EXACT
>QUIT

Here a data file (FROG.DAT) is specified, results will be labeled with the title "Final Analysis of North Fork Frog Study" and results will be written to a file named FLAST.OUT. The statistical procedure called for is an exact, multiresponse (three-variable) permutation procedure (MRPP) on groups of different location. A complete log of all commands entered is kept in a file called BLOSSOM.LOG found in the installed BLOSSOM\LOG folder. Renaming or copying this file after quitting Blossom retains the history of a session.

In this documentation, each command (line) is explained in detail and the complete command syntax for each command is provided.

Commands in Blossom are of two sorts. The first sort is general commands used to specify data, output options, and obtain help. The second sort consists of commands for statistical analyses. In this section, general commands are discussed.

HELP with BLOSSOM Commands

The HELP command gives general help or specific help for Blossom commands. The command line syntax of the HELP command is:

> HELP
> - or -
> HELP <topic>

Where <topic> is the name of the command for which help is sought. The simple HELP command (without a topic specified) lists the topics for which help is available.

For example,

> >HELP

gives Blossom syntax help with a list of all Blossom commands for which syntax help is available, and

> >HELP MRPP

results in help on syntax of the MRPP command.

Additional help is available in the Windows version. The F1 Function Key brings up a Windows Help session for Blossom. The F5 Function Key duplicates the syntax HELP command above. The SHIFT + F5 Function Key brings up the Windows default Web browser with an HTML version of the Blossom User Manual. These help items are also accessible from the Blossom menu bar (Help). The Blossom toolbar help button (with a question mark on it) invokes Windows Help for Blossom.

Alternatively, double-clicking on the installed BLOSSOM\DOCS\BLOSS.HLP file initiates the Windows Help for Blossom and double-clicking on the installed BLOSSOM\DOCS\BLOSSOM.PDF file brings up an Adobe® Acrobat® Reader™ view of the Blossom User Manual (assuming Adobe Acrobat Reader is installed on the computer). The Adobe® Acrobat® Reader™ may be invoked concurrently with either the Console or Windows version Blossom sessions.

USE a Data File

The USE command specifies the data file to be used. The command line syntax of the USE command is:

USE [*data filename*] [/ *variable name list*]

The simple USE statement (with no arguments following on the line) provides a list of files available (Console version) or a Windows file access dialog box. A filename may be specified with the form USE *data filename*. If the file is an ASCII text file and contains no variable names (labels), these should be added after the filename with / *variable name list* (a "slash" followed by variable names in the correct order and number as in the file).

Blossom determines whether the file being read by the USE command is an ASCII text, Data Interchange Format (DIF), S-Plus 2000 (*.), or SYSTAT (SYS or SYD) file. The USE command has two forms for ASCII text data files: USE *filename* and USE *filename* / *variable list*. The first form specifies a labeled and the second an unlabeled data file. (The structure of and differences between these files is described above in the Data Formats section.) For example,

>USE STUDY.DAT

causes Blossom to read the labeled data file STUDY.DAT and provide a list of variables in the file and number of cases read. In this example, the period and file extension need not be entered since "DAT" is the default file extension for Blossom data files. Other file extensions must be supplied explicitly. Data files with no extension are indicated as such by entering the file name followed by a period (e.g., USE DATA.)

To use unlabeled data files specify the variable labels (names) after the name of the file. The command

>USE FIELD1.DAT / GROUP PLOT RESPONS1 RESPONSE2

causes Blossom to read file FIELD1.DAT and assign labels (variable names) GROUP, PLOT, RESPONS1, and RESPONSE2 to the four data variables contained in the file. Labels can be entered in upper or lower case, but is always interpreted by Blossom as upper case. The number of variables in the list following the slash (/) of the USE command must match exactly the number of columns in the data file. Therefore, for example, to analyze only the first four variables in a data file containing six variables (columns), all six variable names must be entered. (Later, a subset of the variables can be specified within the statistical command line).

To read Data Interchange Format (DIF) and SYSTAT (.SYS and .SYD) data files, the entire filename including the extension is entered. Blossom assumes a data file is an ASCII text file with a DAT extension if no extension is provided. Variable names from SYSTAT and Data Interchange Format files are automatically read in by Blossom. The command:

>USE GROUSE.SYS

reads in all variables from the SYSTAT file GROUSE.SYS.

Data in the USEd file are ready for statistical analysis and are available until another USE command is given.

The command:

>USE

(without a filename specification) provides a list of all files in the local folder. The Console version prompts for input of the filename to USE. In the Windows version, this abbreviated command invokes the "Use Data File" dialog box.

In the Console version, a subset of all files is obtained by giving the USE command with a wildcard specification. For example the command:

>USE *.SYS

provides a list of files with a .SYS extension, and the command:

>USE BIRD*.*

lists files with any extension that begin with "BIRD".

In the Windows version a data file can be USEd by selecting "Use/Submit Files | Use Data File" and interacting with the "Use Data File" dialog box. The F2 Function Key or the "Use Dataset" button on the toolbar also invokes this dialog box. A drop-down selection list for "Files of Type" lists (all) Data files (files with extensions DAT, SYS, SYD, or DIF), SYSTAT datasets (files with extensions SYS or SYD), DIF Files (files with extension (DIF), or all files in the local folder.

In the Console version of Blossom, a data file can be USEd by giving the filename as an argument to the CONBLOS.EXE program name at the operating system command prompt. For example, the following Console version session invokes CONBLOS and USEs the BGROUSE.DAT data file:

D:\Blossom\MyData\ElPaso>CONBLOS BGROUSE

```
File being used is BGROUSE.DAT with 21 cases and 3 variables.
The variables are:  DIST, ELEV, SEX
```

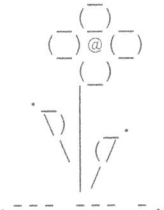

```
BLOSSOM Version C2001.07u
Midcontinent Ecological Science Center
U.S. Geological Survey
4512 McMurry AV
Fort Collins, CO 80525, USA
http://www.mesc.usgs.gov/products/software/blossom.shtml

>
```

In addition, a statistical procedure to be performed can be specified with arguments to the CONBLOS invocation. For example, the following Console version session invokes CONBLOS, USEs the BGROUSE.DAT data file, and runs a multiresponse permutation procedure (MRPP) of variables DIST and ELEV using the grouping variable SEX:

D:\Blossom\MyData\ElPaso>CONBLOS BGROUSE MRPP DIST ELEV * SEX

```
File being used is BGROUSE.DAT with 21 cases and 3 variables.
The variables are:  DIST, ELEV, SEX

         _(_)_
      (_)@(_)
         (_)
    . _      |
     \)  |   _.
      \  | (7
         | /
  .--- --- -.
BLOSSOM Version C2001.07u
Midcontinent Ecological Science Center
U.S. Geological Survey
4512 McMurry AV
Fort Collins, CO 80525, USA
http://www.mesc.usgs.gov/products/software/blossom.shtml



              Multi-Response Permutation Procedure (MRPP)

  Data Used
            Data File: BGROUSE.DAT
       Grouping Variable: SEX
      Response Variables: DIST, ELEV

Specification of Analysis
   Number of observations: 21
         Number of groups: 2
         Distance exponent: 1.00000000000000
         Weighting factor: n(I)/sum(n(I)) = C(I) = 1

Group Summary
   Group Value                    Group Size  Group Distance
    3.00000000000000                      9   1.07214652525827
    4.00000000000000                     12   1.39643892970427

Variable Commensuration Summary
     Variable Name        Average Distance (Euclidean if V=1)
     DIST                      9264.76190476191
     ELEV                      279.228571428571
```

```
Results
   Delta Observed  =  1.25745647065599
   Delta Expected  =  1.51256336315532
   Delta Variance  =  0.270618755524092E-002
   Delta Skewness  =  -2.09758982732985

              Standardized test statistic = -4.90391852737653
                 Probability (Pearson Type III) of a
                    smaller or equal delta = 0.298316800990588E-002

Output was appended to file "BGROUSE.OUT"

>
```

ECHO Data On Input or Results on Output

The ECHO command is used to control extent of information displayed to the output window (or console screen). The command line syntax of the ECHO command is:

> ECHO DATA=ON|OFF OUTPUT=ON|OFF
> - or -
> ECHO DEFAULT

ECHO can be used with either the DATA or OUTPUT specifier, or both, or with the DEFAULT specifier.

Echoing data on input allows inspection of data values read into Blossom when a file is accessed with the USE command. Turning this echo off reduces the amount of text scrolled in the output window (console screen). ECHO DATA=ON causes data values to be displayed, and ECHO DATA=OFF stops this display. The default is OFF, as normally a user has no need to re-inspect data values.

Writing statistical results to the output window (console screen) is the default for Blossom. In some situations, such as the processing of large submit files (Blossom command files) with the SUBMIT command, the extensive writing of output to the screen may increase program runtime. Turning off the echo of output decreases program runtime by reducing text written to the screen. All statistical results are always written to the Blossom output file, so there is no problem recovering results from such a "batch-mode" session.

When engaged in an interactive session with Blossom, a user normally prefers to view statistical results immediately, so the default mode is ECHO OUTPUT=ON. ECHO OUTPUT=OFF stops screen output of these results. In any case, all statistical results are written to an output file.

ECHO DEFAULT sets the echo modes for data and output results to Blossom default values and is the same as ECHO DATA=OFF OUTPUT=ON.

For example,

>ECHO DATA=ON

tells Blossom to show values from the USEd data file, and

>ECHO OUTPUT=OFF DATA=OFF

tells Blossom to stop screen output of statistical results and to not display data values as they are read. The command:

>ECHO DEFAULT

sets ECHO values to their default, which is the same as.

>ECHO DATA=ON|OFF OUTPUT=ON|OFF

SHELL to DOS

The SHELL command works only in the Console version. The command line syntax of the SHELL command is simply:

SHELL

The SHELL command allows the user to temporarily "return" to the operating system. There the user can issue operating system commands or run other programs. The command:

>SHELL

puts the user at the operating system command prompt. The command EXIT followed by ENTER returns to Blossom. The SHELL command is useful for editing data files (shell out of Blossom, edit the file, save, and return to Blossom) and viewing output files or the history of the current session in the installed BLOSSOM\LOG\BLOSSOM.LOG file.

This command is a relict of the old DOS version of Blossom where there was no multitasking capability. It is best to use Windows facilities to accomplish these other tasks while Blossom runs in its own window.

SAVE a Data File as Labeled

The SAVE command saves currently used data from an unlabeled file into a labeled data file. The command line syntax of the SAVE command is:

SAVE *labeled filename*

where *labeled filename* is the name of the file to create for saving labeled data. For example,

>SAVE DATA2

saves a labeled data file, DATA2.DAT, with the labels being those specified on the previously entered USE command. This command is useful for changing an unlabeled data file into a labeled one, which in subsequent sessions saves typing the variable list with the USE command. The name specified in the SAVE command must be different from that of the file in use and may include a file extension.

OUTPUT Results to Specified File

The OUTPUT command specifies the name of a file to which results of statistical analyses are to be written. The command syntax of the OUTPUT command is:

OUTPUT *filename*

where *filename* is the name of the file to which Blossom is to direct statistical results. For example,

>OUTPUT FISH.OUT

places results to the file FISH.OUT for all analyses specified until the session is terminated or another OUTPUT command is given. If the output file already exists, results are appended to it; it is not overwritten. If another output file is in use, it is "closed" and the new file becomes the output file.

If no OUTPUT command is given, results are written to a default output file. The name of the default output file is the same as the file given with the USE command, but with an "OUT" file extension. Results are appended to an already existing output file.

If an OUTPUT command is given to name an output file, that file is the output file for Blossom and subsequent USE commands does change the output file name.

Results, besides being written to an output file, are written to the screen as well (cf. ECHO OUTPUT=OFF command) unless the TERSE option is selected.

The options TERSE and VERBOSE are provided to turn on and off a terse formatting of the saved output file. The TERSE option also automatically assumes not to ECHO the output to the screen. The TERSE option is primarily intended to provide a very concise formatting of relevant output for multiple runs of the same commands from a submit file (e.g., as would be done in a large simulation experiment). The format of the TERSE output is one line per command with the first column being the Blossom command executed (e.g., LAD), the second column is the USEd file name, and subsequent columns are relevant parameter estimates, test statistics and *P*-values as appropriate to the command. Column variable names are not provided in the output file so it is important for the user of the TERSE option to know and label these columns appropriately.

>OUTPUT FISH.OUT/TERSE

turns on the terse output which remains in effect until a

>OUTPUT FISH.OUT/VERBOSE

command is given. The default assumes VERBOSE.

TITLE for the Output of Results

The TITLE command gives the opportunity to specify text that is written at the beginning of each set of results from statistical procedures. The command line syntax of the TITLE command is:

TITLE *text of title*

where *text of title* is the text to be used as a heading of statistical results. Entering a new TITLE command changes the title. The entry

>TITLE First Analysis of Storm River Data - 2 groups

places the indicated text at the beginning of each subsequent set of statistical results written by Blossom. A TITLE command with no text specified causes no title to be written and serves to cancel a previous TITLE command.

DATE to Screen and Output File

The DATE command writes the current date to the screen and current output file if it exists. It is useful for dating results and can be used to time procedures if issued before and after a statistical command. The command line syntax for the DATE command is simply:

DATE

For example, the command:

>DATE

immediately writes the current date and time to the screen and output file.

RANDOM Specify Random Number Generator

2. Use the RANDOM command to specify the pseudo-random number generator that Blossom should use. The syntax for the RANDOM command is:

RANDOM = DEFAULT
- or -
RANDOM = MT
- or -
RANDOM

By default Blossom uses a multiplicative congruential algorithm. To invoke the Mersenne Twister algorithm, use the command RANDOM=MT. To reinstate the default algorithm, use the command RANDOM=DEFAULT. If the command is given as simply RANDOM, Blossom will display the syntax for the RANDOM statement and display the current random number generator.

CD Change Directory (Move to New Data Path)

The CD command is used to change the current Data Path (file folder) where Blossom is operating.

In the Windows version of Blossom, after installation the Data Path is the Installed BLOSSOM\SAMPLES directory. A record is kept by Blossom of subsequent changes of the Data Path when a CD, USE, or SUBMIT command is used. New Windows Blossom sessions will begin in the last used Data Path.

In the Console version of Blossom, the initial Data Path is always in the current working directory where the program is invoked. Subsequent CD commands can be used to navigate the file system. A record is kept of the location of the Data Path. Subsequent Windows Blossom sessions will begin in the last used Data Path (including Console sessions).

The following CD command changes to the \DATA\OSPREY2K directory on the current drive:

```
>CD \DATA\OSPREY2K
```

The following command moves up one level in the directory structure:

```
>CD ..
```

The following command moves to the K: drive:

```
>CD K:
```

The following command moves to the TEAL subdirectory (below the current Data Path:

```
>CD TEAL
```

STATUS of BLOSSOM Session

The STATUS command gives information on the current Blossom session. The command line syntax for the STATUS command is simply:

```
STATUS
```

Included in the status report is the name of the current data file being USEd, the number of cases, the number of variables and variable names, the names of the SAVE and OUTPUT files (if any), the current TITLE text, and the most recent LAD and HYPOTHESIS commands. If a USE command has not yet been given to specify a data file, a warning message is displayed. Also included is whether the OUTPUT is VERBOSE or TERSE and the random number generator currently in use.

Type

```
>STATUS
```

to see a the information for the current Blossom session

SUBMIT a Command File

The SUBMIT command causes Blossom to read commands from an input file rather than the command line. The command line syntax for the SUBMIT command is:

```
SUBMIT filename
```

where *filename* is the filename of the file containing Blossom commands to be executed.

In this way, "programs" can be submitted to Blossom for carrying out long or repetitive analyses or to exactly repeat an analysis already performed. An ASCII text file containing any valid Blossom commands can be submitted. The STATUS and SHELL commands are of little use with the SUBMIT command, however the comment command (' or ", see below) is useful for documenting analyses called for in the submit file. It is possible to copy the BLOSSOM.LOG to another file, edit it, save it as a submit file, and submit the modified file. The command:

> >SUBMIT SUBWAY

causes Blossom to process the set of commands in submit command file TEST1. If the submitted file has other than the "SUB" file extension then its complete name must be specified. If the file has the default extension (SUB), it need not be specified.

In the Console version, the command SUBMIT without a file specification produces a list of files with the SUB extension. The desired file can then be specified.

In the Windows version a "Submit Command File" dialog box can be invoked from the Blossom menu bar selection "Use/Submit Files | Submit Command File", or from the"Submit Command File" button on the toolbar, or with the SHIFT + F2 Function Key. A drop-down selection listing of submit files (files with the extension SUB) or all files in the local folder can be obtained.

Advanced SUBMIT Operations with Program Arguments and DOS Batch Files

Both the Windows and Console versions of Blossom can be invoked from a DOS Batch file. If the last command of a submitted command file is QUIT, control returns to the batch file for further processing.

For example, the two Blossom command submit files:

```
' File: sub1.sub
output subtest1
use bgrouse
mrpp dist elev * sex
quit
```

and

```
' File: sub2.sub
output subtest1
use mrbp.dat
mrpp spp1 spp2 spp3 * trtmt * block
quit
```

can be invoked from a batch file called BATWIN.BAT:

```
REM File: batwin.bat
REM Run two blossom submit files
REM Windows version
blossom submit sub1
blossom submit sub2
```

When the BATWIN.BAT is invoked, Blossom starts and the SUB1.SUB file is submitted for processing. When that is finished, the SUB2.SUB file is submitted for processing. Control is then returned to the system. The resultant SUBTEST1.OUT output file looks like this:

```
====================================================================

                    Multi-Response Permutation Procedure (MRPP)

    Data Used
                        Data File: BGROUSE.DAT
           Grouping Variable: SEX
          Response Variables: DIST, ELEV

    Specification of Analysis
          Number of observations: 21
                Number of groups: 2
                Distance exponent: 1.00000000000000
                Weighting factor: n(I)/sum(n(I)) = C(I) = 1

    Group Summary
        Group Value                    Group Size   Group Distance
        3.00000000000000                       9    1.07214652525827
        4.00000000000000                      12    1.39643892970427

    Variable Commensuration Summary
          Variable Name            Average Distance (Euclidean if V=1)
          DIST                     9264.76190476191
          ELEV                     279.228571428571

    Results
        Delta Observed = 1.25745647065599
        Delta Expected = 1.51256336315532
        Delta Variance = 0.270618755524093E-002
        Delta Skewness = -2.09758982733399

                Standardized test statistic = -4.90391852737653
                    Probability (Pearson Type III) of a
                    smaller or equal delta = 0.298316800991671E-002

    =================================================================

    Multi-Response Permutation Procedure for Blocked Data (MRBP)

    Data Used
                        Data file: MRBP.DAT
           Grouping Variable: TRTMT
           Blocking Variable: BLOCK
          Response Variables: SPP1, SPP2, SPP3

    Specification of Analysis
          Number of observations: 18
                Number of groups: 6
                Number of blocks: 3
                Distance exponent: 1.00000000000000

    Group Summary
        Group Value                    Group Size
        1.00000000000000                       3
```

```
        2.00000000000000                        3
        3.00000000000000                        3
        4.00000000000000                        3
        5.00000000000000                        3
        6.00000000000000                        3

    Block Alignment Summary
        Block Value                  Variable Name              Alignment Value
        1.00000000000000             SPP1                       6.50000000000000
                                          SPP2
                             3.16500000000000
                                          SPP3
                             2.17000000000000
        2.00000000000000             SPP1                       9.91500000000000
                                     SPP2                       1.16500000000000
                                     SPP3                       2.66500000000000
        3.00000000000000             SPP1                       6.25000000000000
                                     SPP2                       1.91500000000000
                                     SPP3                       2.41500000000000
    Variable Commensuration Summary
        Variable Name                Average Euclidean Distance
        SPP1                         7.60150326797386
        SPP2                         3.10692810457516
        SPP3                         0.900588235294119

    Results
        Delta Observed = 1.78519097155486
        Delta Expected = 1.98049119623354
        Delta Variance = 0.209317316371645E-001
        Delta Skewness = -0.389741935641221

    Agreement measure among blocks = 0.986120135500246E-001
        Standardized test statistic = -1.34989554442147
            Probability (Pearson Type III) of a
                smaller or equal delta = 0.949929802101351E-001
```

In a similar fashion, the DOS batch file BATCON.BAT

```
    REM File: batcon.bat
    REM Run two blossom submit files
    REM Console version
    conblos submit sub1
    conblos submit sub2
```

invokes the Console version of Blossom using the same submitted command files as above and produces identical results.

If the last command in the submitted command file is QUIT, control returns to the operating system prompt. With this in mind, a DOS Batch file can be created that invokes several submit files in succession. With the DOS change directory (CD) commands, a session could process several folders of data by running one Batch file.

Comments in Log, Data and Submit Files (The Quote Command)

A comment is indicated by a single or double quotation mark (' or ") as the first non-blank character of a line. The command line syntax of the comment is:

```
' text of comment
- or -
" text of comment
```

where *text of comment* is the text of the comment to be inserted in the Blossom history (BLOSSOM.LOG file). Comments can be entered at the Blossom command line, in which case the comment is added to the BLOSSOM.LOG file to help document a session. Comments can also be used within ASCII text data or submit files to indicate what data are used, what the variable names mean, and what analyses are being called for. Blossom skips over comment lines in data or submit files. Comments are useful for annotating steps of analysis throughout a session. For example, entering:

```
>'now calculate a quadratic LAD regression on ht versus age
```

writes the comment line to the current session's log file.

A data file with comments might look like this:

```
'   Spatial coordinates of young and old birds
'   data collected summer 1989
GROUPT X_COORD Y_COORD
' begin group 1 = young
1       4       5
1       3       4
1       4       3
' begin group 2 = old
2       2       3
2       2       2
2       3       2
2       3       1
```

Contrast this comment function (which writes to the BLOSSOM.LOG file) with the NOTE command below (which writes to the OUTPUT file).

NOTE to Output File

The NOTE command writes the contents of the command line after "NOTE" to the OUTPUT file. The command line syntax of the NOTE command is:

```
NOTE text of note
```

where *text of note* is the text to be included as a note in the OUTPUT file.

An OUTPUT file must be open for a note to be written, i.e., a USE or OUTPUT command must have been given in the Blossom session prior to the NOTE command for a note to be written. This command is useful to annotate the OUTPUT file to document a session.

>NOTE The data for this MRPP is from Uncompagre for May, 2000

Contrast the NOTE command (which writes to the OUTPUT file) with the ' or " (comment) command above (which sends a comment to the BLOSSOM.LOG file).

QUIT BLOSSOM Session

The QUIT command ceases execution of the Blossom program. The command line syntax of the QUIT command is simply:

QUIT
- or -
QU

If the console version of Blossom is running, QUIT returns the user to the operating system prompt.

Simply type

>QUIT

to quit the Blossom session.

In the Windows version, you can also quit the Blossom session by using the "File | Exit" menu selection, by clicking on the Windows "X" (Close) button on the top right of the Blossom window title bar, or by entering the ALT + F4 key.

Windows Version Specific Commands and Functions

The Windows version of Blossom has some Windows graphical user interface features.

The Windows version of Blossom has a menu with five main menu selections. Under these are submenu selections. The Blossom submenu selections nearly all invoke equivalent general program functions as discussed in the General Program Functions section. Some functions are unique to the Blossom menu and toolbar and these are explicitly discussed here. The function invoked by each selection is related here.

The Blossom toolbar consists of buttons below the Blossom menu. These duplicate some of the menu functions.

In addition, there are Function Key and keyboard shortcuts (key combinations) that invoke some Blossom features.

CLS to Clear Blossom Output Window

In the Windows version of Blossom the CLS command clears (erases) the contents of the output window (immediately above the "Blossom Command>" entry field). The command line syntax (from the "Blossom Command>" entry field is simply:

CLS

This is useful to eliminate any previous output before printing or saving contents of the Output Screen.

Windows Blossom Menu

The Windows version of Blossom has a menu bar with five main menu selections. Under these are submenu selections. The Blossom submenu selections nearly all invoke equivalent general program functions as discussed in the General Program Functions section. Some functions are unique to the Blossom menu and toolbar and these are explicitly discussed here. The function invoked by each selection is related here.

File

Print
The "File | Print" menu selection prints the contents of the Blossom Windows version output window to the Windows printer. A "Page Setup" dialog box appears and the user can select options for printer output, including selecting the printer and printer properties. This function may be invoked using the CONTROL + P key combination.

Print Selection
The "File | Print Selection" menu selection invokes the same dialog box as the "File | Print" menu selection, but only the text selected (highlighted) by the user in the Blossom output window is sent to the Windows printer.

Exit
The "File | Exit" menu selection stops the Blossom session and stops the program execution. This is the same as invoking the QUIT command or by clicking the "X" (Close) button on far right of the Blossom Windows version program title bar. The standard Windows ALT + F4 key combination also causes the program to quit.

Edit

Copy

The "Edit | Copy" menu selection copies the text selected (highlighted) by the user in the Blossom Windows version output window into the Windows Clipboard. This text can then be pasted into other programs. This function may be invoked using the CONTROL + C key combination if the input cursor focus is in the Blossom Windows version output window.

Select All

The "Edit | Select All" menu selection selects (highlights) all the text in the Blossom Windows version output window. The selected text subsequently may be copied into the Clipboard. This function may be invoked using the CONTROL + A key combination if the input cursor focus is in the Blossom Windows version output window.

Search

Find

The "Search | Find" menu selection opens a "Find" dialog box. The user can enter text for which to search from within the Blossom Windows version output window. This function may be invoked using the CONTROL + F key combination.

Find Next

The "Search | Find Next" menu selection searches for the next occurrence of the text specified in the "Search | Find" selection. A (text) Find search within the Blossom output window must have been initiated. Once a search is underway, this function can be invoked using the F3 Function Key.

Use/Submit File

Use Data File

The "Use/Submit File | Use Data File" menu selection invokes the "Use Data File" dialog box as discussed in the USE command above. This function can be invoked using the F2 Function Key or by clicking on the "Use Dataset" button on the toolbar.

Submit Command File

The "Use/Submit File | Submit Command File" menu selection invokes the "Submit Command File" dialog box as discussed in the SUBMIT command above. This function can be invoked using the SHIFT + F2 Function Key or by clicking on the "Submit Command.File" button on the toolbar.

Help

There are several Help selections available. Make a selection based on your needs.

BLOSSOM Help
The "Help | Blossom Help" menu selection invokes a Window Help session with a Blossom Help file. Normal Windows Help functions are available including Find and searching for Help Topics within the Blossom Help file. This function may be invoked with the F1 Function Key or the "Help" button on the toolbar.

BLOSSOM Syntax Help
The "Help | Blossom" Syntax Help menu selection sends a list of commands for which there is syntax help, just as the HELP command discussed above. The user can use the "HELP <topic>" command line to obtain syntax help on a topic. This function may be invoked with the F5 Function Key.

User Manual (Local Browser)
The "Help | User Manual (Local Browser)" menu selection invokes the default Web browser on the user's computer and opens an HTML version of the Blossom User Manual. This function may be invoked with the SHIFT + F5 Function Key.

About BLOSSOM
The "Help | About Blossom" menu selection displays a small dialog box with information about the Blossom Windows version.

WWW: BLOSSOM Updates on Web
The "Help | WWW: Blossom Updates on Web" menu selection invokes the default Web browser on the user's computer and attempts to open the URL http://www.fort.usgs.gov/products/software/blossom/blossom.asp and display the latest Blossom Web page. Any updates to Blossom programs can be found there.

WWW: FORT USGS Homepage on Web
The "Help | WWW: FORT USGS Homepage on Web" menu selection invokes the default Web browser on the user's computer and attempts to open the URL http://www.fort.usgs.gov/ and display the Fort Collins Science Center, U.S. Geological Survey Homepage. This is the institution where Blossom was developed.

Windows Blossom Toolbar

The Blossom toolbar consists of buttons below the Blossom menu. These buttons duplicate some of the menu functions.

Print Button

The "Print" button of the Blossom toolbar prints output window contents to the Windows printer. It has the same function as the "File | Print" menu selection discussed above and may be invoked with key combination CONTROL + P.

Find Button
The "Find" button of the Blossom toolbar has the same function as the"Search | Find" menu selection discussed above. It is used to search for text within the Blossom Windows version output window. It may be invoked with the CONTROL + F key combination.

Copy Button
The "Copy" button of the Blossom toolbar copies selected (highlighted) text from the Windows version output window to the Clipboard. It has the same function as the "Edit | Copy" menu selection and may be invoked with the CONTROL + C key combination if the input cursor focus is in the Blossom Windows version output window.

Use Dataset Button
The "Use Dataset" button of the Blossom toolbar has the same function as the simple "USE" command from the command line and the "Use/Submit Files | Use Data File" menu selection and the F2 Function Key. It invokes a ">Use Data File" dialog box as discussed with the USE command above.

Submit Command File Button
The "Submit Command File" button of the Blossom toolbar has the same function as the simple "SUBMIT" command from the command line and the "Use/Submit Files | Submit Command File" menu selection and the SHIFT + F2 Function Key. It invokes a "Submit Command File" dialog box as discussed with the SUBMIT command above.

Blossom Help Button
The "BLOSSOM Help" button has the same function as the "Help | Blossom Help" menu selection and the F1 Function Key. It invokes a Windows Help session with the Blossom Windows Help file.

Function Keys and Keyboard Shortcuts in Windows Blossom

The Function Keys and Keyboard Shortcuts perform the same functions as the menu selections and Command line entries (except for the F4 Function Key, which has a unique function not accessible from other sources). Standard Windows editing key combinations operate within the "Blossom Command>" entry field.

F1 Function Key - Blossom Windows Help

The F1 Function Key invokes a Windows Help session with Blossom Help. The same function can be accessed from the "Help | Blossom Help" menu selection and the "BLOSSOM Help" button on the toolbar.

F2 Function Key - Use Data File
The F2 Function Key invokes the "Use Data File" dialog box as discussed in the USE command above. This function can be invoked using the "Use/Submit File | Use Data File" menu selection or by clicking on the "Use Dataset" button on the toolbar.

SHIFT + F2 Function Key - Submit Command File
The SHIFT + F2 Function Key invokes the "Submit Command File" dialog box as discussed in the SUBMIT command above. This function can be invoked using the "Use/Submit File | Submit Command File" menu selection or by clicking on the "Submit Command.File>" button on the toolbar.

F3 Function Key - Find Next
The F3 Function Key searches for the next occurrence of the text specified in the "Search | Find" selection. A (text) Find search within the Blossom output window must have been initiated. Once a search is underway, this function can be invoked using the "Search | Find Next" menu selection.

F4 Function Key - Command History Popup
When the input cursor is focused within the "Blossom Command>" entry field, the F4 Function Key invokes a popup list selection box with a list of up to 100 previous commands the user has entered during the current Blossom session. Clicking on (selecting) a command recalls it to the "Blossom Command>" entry field where it may be edited or accepted and then entered (press the ENTER key). The F4 Function Key is the only way to invoke this operation.

ALT + F4 Function Key - Quit Blossom Session
The standard Windows ALT + F4 Function Key ceases the Blossom session and stops the program execution. This is the same as invoking the QUIT command or by clicking the "X" (Close) button on far right of the Blossom Windows version program title bar. The "File | Exit" menu selection also causes the program to quit.

F5 Function Key - BLSSOM Syntax Help
The F5 Function Key sends a list of commands for which there is syntax help, just as the HELP command discussed above. The user can use the "HELP *topic*" command line to obtain syntax help on a topic. This function may be invoked with the "Help | Blossom" Syntax Help menu selection.

SHIFT + F5 Function Key - User Manual

The SHIFT + F5 Function Key invokes the default Web browser on the user's computer and opens an HTML version of the Blossom User Manual. This function may be invoked with the "Help | User Manual (Local Browser)" menu selection.

F10 Function Key - Access Menu Bar

The standard Windows F10 Function Key function accesses the Blossom program menu bar.

CTRL + A Key Combination - Select All

With the input cursor focused in the Blossom Windows version output window, the CTRL + A key combination selects (highlights) all text in that window. This text may then be copied to the Windows Clipboard.

CTRL + C Key Combination - Copy (to Clipboard)

With the input cursor focused in the Blossom Windows version output window, the CTRL + C key combination copies selected (highlighted) text in that window to the Windows Clipboard.

CTRL + F Key Combination - Find

The CTRL + F key combination opens a "Find" dialog box.The user can enter text for which to search from within the Blossom Windows version output window. This function also may be invoked using the "Search | Find" menu selection.

CTRL + P Key Combination - Print

The CTRL + P key combination prints the contents of the Blossom Windows version output window to the Windows printer. A "Page Setup" dialog box appears and the user can select options for printer output, including selecting the printer and printer properties. This function may be invoked using the "File | Print" menu selection.

Statistical Commands

Blossom currently has six statistical commands, MRPP, SP, MEDQ, LAD, OLS, and COV. The MRPP command can specify one of three multiresponse permutation procedures.

1) Multiresponse permutation procedures (MRPP)

2) Multiresponse randomized block permutation procedures (MRBP)

3) Permutation tests for matched pairs (PTMP)

These procedures (MRPP, MRBP, and PTMP) are distribution-free techniques for making inferences about grouped data. Their advantages over many classical techniques include the ability to select an analysis space commensurate with the geometry of the data as perceived by the investigator. Several classical univariate and multivariate parametric and rank tests can be emulated with these procedures as well. The simplest MRPP analysis is for data consisting of two or more observations on objects in two or more groups. The MRBP and PTMP variants are for similar data that are blocked or paired.

Since the MRPP command can emulate so many different statistical tests, the specification of the command line can be quite complex. However, Blossom uses default values, which for routine analysis makes the command easy to use.

The MEDQ command calculates univariate or multivariate medians and distance quantiles either by groups specified by a grouping variable or for the entire data file being used. Options allow you to specify quantiles to report that differ from the default quantiles.

The SP command calculates the multiresponse sequence procedure to test for first-order autoregressive patterns (serial dependency). The default value produces an analysis in Euclidean space. A sequencing variable that determines the order of the data can be selected or Blossom assumes by default that the order in the file is the sequential order of interest.

The LAD command estimates a least absolute deviation regression or an optional quantile regression. The model specified in the LAD command line is considered the full parameter alternative model for hypothesis tests. The associated command, HYPOTHESIS, can be used to specify a reduced parameter null model that is tested against the model specified by the LAD command.

The OLS command estimates an ordinary least squares regression. It has an associated HYPOTHESIS command that performs a similar function in testing hypotheses as the associated HYPOTHESIS command does with the LAD command.

The COV command provides for tests of g-sample empirical coverage tests it used with a grouping variable and related goodness-of-fit tests if specified without a grouping variable.

The MRPP variants, MRSP, LAD, OLS, and COV are discussed in turn. MEDQ is discussed with MRPP as it provides descriptive estimates that are useful for interpreting results of hypothesis tests with MRPP.

Multiresponse Permutation Procedure (MRPP)

MRPP is best introduced with an example. The following is a bivariate example adapted from Biondini et al. (1985). A similar example is found in Zimmerman et al. (1985), Biondini et al. (1988), and a univariate example is given in Slauson (1988).

In Figure 1 the values of two variables, x and y, are shown for seven observations in two groups, A and B.

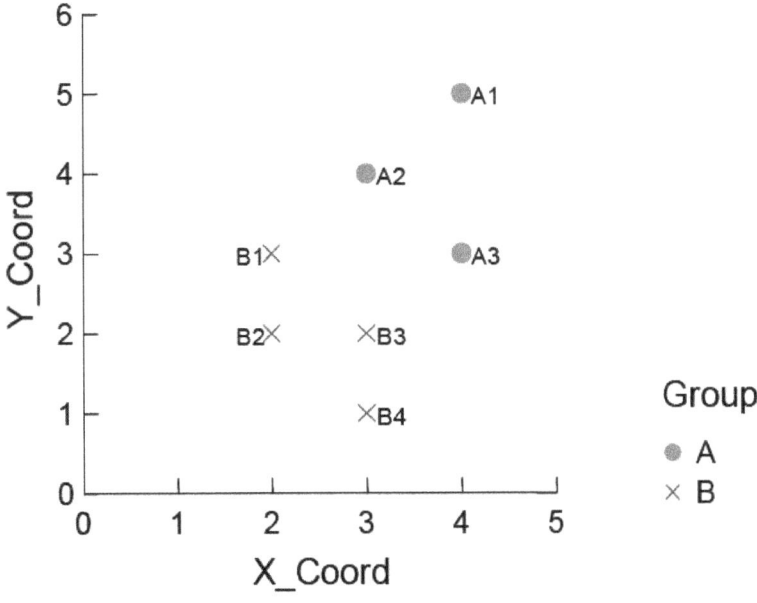

Figure 1. The observed sample for 2 groups with bivariate response Y_Coord and X_Coord.

The objects in groups A and B seem to be clustered or concentrated in different parts of the x-y plane representing the two response (measured) variables x and y. One way to determine if the two groups are so clustered is to measure or calculate the distances between all pairs of members of each group and calculate an average distance for each group (A = 1.609, B = 1.344). If group members are clustered together, then the intragroup average distances will be small compared to cases where the group members are spread out and overlap more with other groups. For example, Figure 2 shows the same data except that the groups that observations A3 and B2

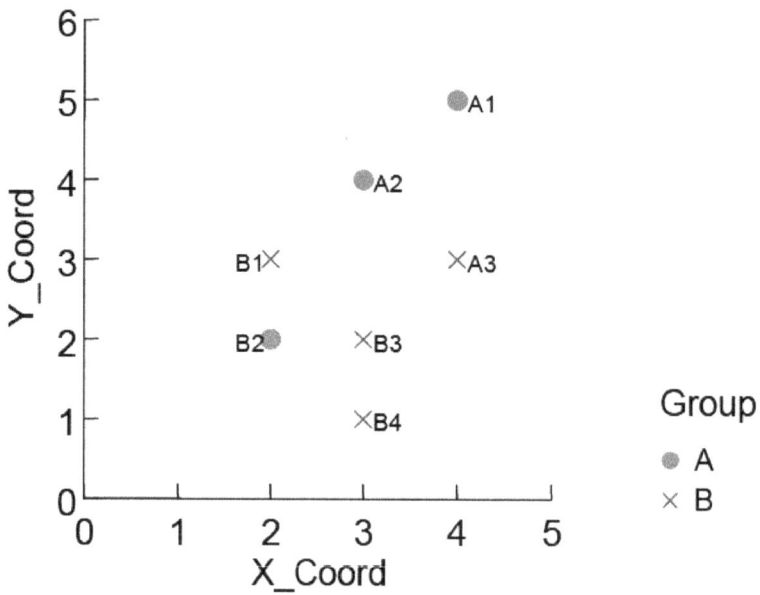

Figure 2. One of the possible other 34 permutations of the data in Figure 1.

belong to are switched. In this case the intragroup average distances will be greater than for the case first shown above (A = 2.419, B = 1.717).

The strategy of MRPP is to compare the observed intragroup average distances with the average distances that would have resulted from all the other possible combinations of the data under the null hypothesis. The test statistic, usually symbolized with a lower case delta, δ, is the average of the observed intragroup distances weighted by relative group size, 3/7 and 4/7 in this case. The observed delta (δ_{obs}) is compared to the possible deltas (δ) resulting from every permutation of the above 7 points into 2 groups of 3 and 4 members. If the hypothesis that the two groups are not different (the null hypothesis) is true, then each of the possible assignments (permutations) is equally likely. In this example there are 35 permutations possible, each with a 1/35 (1/35 = 0.0286) chance of occurring. Here are the Blossom commands to read in the data file, EXAMPLE1.DAT, and compute the MRPP results.

```
>USE EXAMPLE1.DAT / GROUP X_COORD Y_COORD
>MRPP X_COORD Y_COORD * GROUP / NOCOM EXACT
```

X_COORD and Y_COORD are the 2 response variables, GROUP is the grouping variable, and the exact version of MRPP is chosen since this is such a small sample. NOCOM signifies that no multivariate commensuration is desired. Blossom by default will commensurate multiple variables by the average Euclidean distance for each variable ignoring group structure. Think of this as similar to the usual parametric approach of standardizing variables to unit variance (average squared Euclidean distance).

Here are the results:

```
                 Exact Multi-Response Permutation Procedure (EMRPP)

 Data Used
                      Data File: Example1.dat
   Grouping Variable: GROUP
 Response Variables: X_COORD, Y_COORD

 Specification of Analysis
    Number of observations: 7
            Number of groups: 2
          Distance exponent: 1.00000000000000
           Weighting factor: n(I)/sum(n(I)) = C(I) = 1

 Group Summary
   Group Value                    Group Size
   1.00000000000000                     3
   2.00000000000000                     4

 Variables are not commensurated

 Results
                          Observed delta = 1.45782245613148
 Probability (Exact) of a smaller or equal delta = 0.285714285714286E-001
```

The probability value (*P*-value) is 0.0286 which means that the observed delta was the smallest among the 35 possible deltas.

Use the EXACT option for MRPP with caution for it can take a long time if the sample sizes are greater than about 20, depending on the computer.

By default MRPP does not compute exact probabilities but uses an approximation of the exact distribution of the test statistic (δ) to estimate the *P*-value. The default approximation is based on the first three exact moments (mean, variance, and skewness) of the permutation distribution evaluated as a Pearson type III distribution (Berry and Mielke 1983, Iyer et al. 1983, Mielke and Berry 2001). The moments approximation avoids the simulation error associated with Monte Carlo resampling tests (Mielke and Berry 1982; Berry and Mielke 1985). However, we offer the option of approximating the permutation distribution of the test statistic with a Monte Carlo resampling procedure with the option NPERM. By default NPERM uses 5,000 (4,999 + observed delta) random samples to approximate the permutation distribution but the user may specify any desirable number of resamples, e.g., NPERM = 10000. Most examples we've encountered yield similar *P*-values from the Monte Carlo resampling and Pearson type III distribution approximations, but it is possible for the Monte Carlo resampling approximation to yield better estimates for some problems, e.g., with a large number of discrete values clumped in some region of the data space or if interest is in upper tail probabilities (e.g., $P > 0.90$) associated with detecting regularity of spatial data distributions. Further investigation of these properties is an open area for research.

The next example shows how to emulate a 2-sample t-test with MRPP. Consider the data for two groups in Figure 3 (from Mielke 1986). The single response variable is represented on the horizontal axis and the number of observation on the vertical.

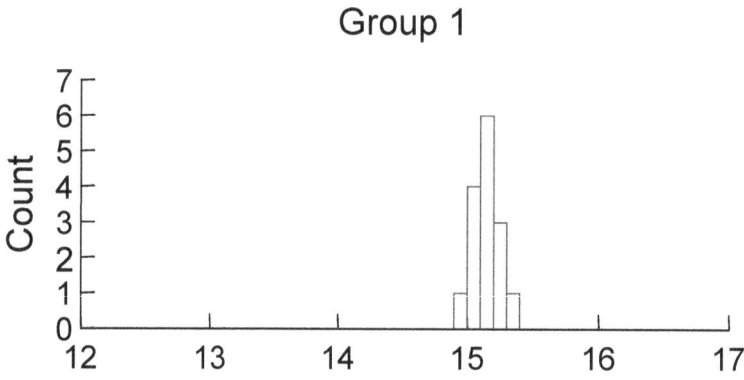

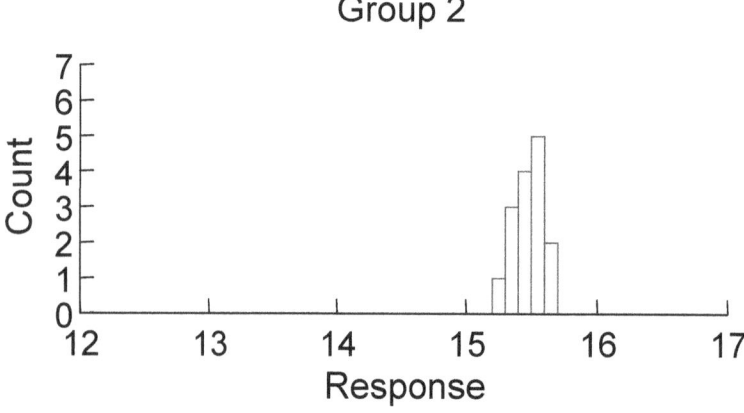

Figure 3. Two-group example from Mielke 1986 (no outliers)

Group 1 (median = 15.10, mean = 15.09) and 2 (median = 15.40, mean = 15.42) appear to differ slightly (0.3) in central tendency. To test for equality of means with the t-test, USE the data file EXAMPLE3.DAT, specify a title if desired, and enter the following MRPP command.

>MRPP RESPONSE * GROUP / V=2 C=2

The V = 2 option causes MRPP to compute squared Euclidean distances (V = 1 is the default value and specifies Euclidean distance). The C = # option specifies how the intragroup distances are to be averaged. If C = 2 is specified, then the analysis mimics the classical parametric t-test, where the group distances are weighted by the relative degrees of freedom. If C = 1 then the intragroup distances are weighted by relative group size, then averaged to arrive at delta. This is the default value. In this example, since the group sizes are equal, the choice of C does not matter. In general choose C = 2 and V = 2 to calculate a test that mimics the classical parametric t- and F-tests for univariate data and Hotelling's T-square or MANOVA for multivariate data. Here are the results of the above MRPP command:

```
                Multi-Response Permutation Procedure (MRPP)

 Data Used
                     Data File: Example3.dat
        Grouping Variable: GROUP
     Response Variables: RESPONSE

 Specification of Analysis
     Number of observations: 30
            Number of groups: 2
          Distance exponent: 2.00000000000000
          Weighting factor: (n(I)-1)/sum(n(I)-1) = C(I) = 2

 Group Summary
     Group Value                   Group Size  Group Distance
     1.00000000000000                      15  0.213333333333333E-001
     2.00000000000000                      15  0.270476190476190E-001

 Results
     Delta Observed = 0.241904761904761E-001
     Delta Expected = 0.808275862068974E-001
     Delta Variance = 0.156341252315437E-004
     Delta Skewness = -2.56497266493768

             Standardized test statistic = -14.3239993158952
                 Probability (Pearson Type III) of a
                   smaller or equal delta = 0.192580769475062E-005
```

The very small P-value (0.0000019) indicates that these two samples are unlikely to come from populations with the same mean, i.e., they are different. The two sample t-test based on normal theory also gives a very low P-value for these data (P < 0.000001).

Now consider the same data, but with one difference, viz, a change in one of the 30 data values (Fig. 4).

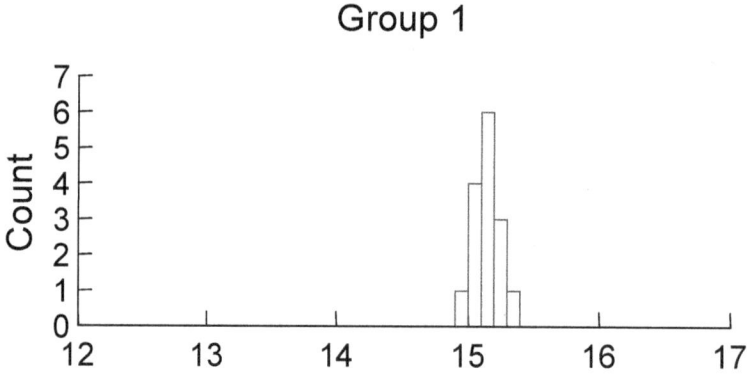

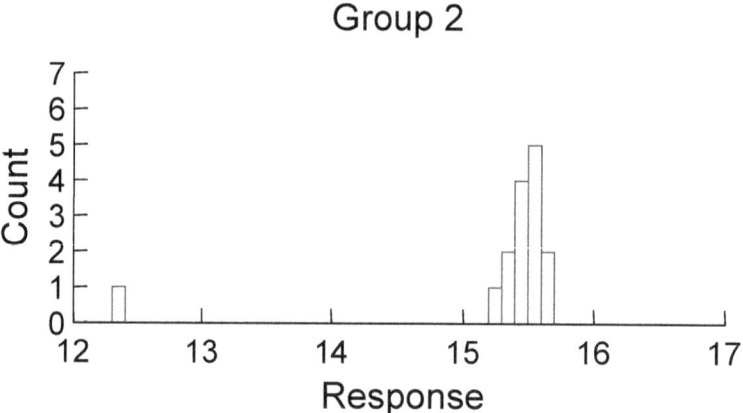

Figure 4. Two-group example from Mielke 1986 (one outlier in Group 2)

To compare these samples USE the file EXAMPLE4.DAT and issue the following MRPP command.

>MRPP RESPONSE * GROUP / V=2 C=2

Here are the results:

```
             Multi-Response Permutation Procedure (MRPP)

Data Used
              Data File: EXAMPLE4.DAT
     Grouping Variable: GROUP
    Response Variables: RESPONSE

Specification of Analysis
```

```
     Number of observations: 30
           Number of groups: 2
          Distance exponent: 2.00000000000000
          Weighting factor: (n(I)-1)/sum(n(I)-1) = C(I) = 2

Group Summary
   Group Value                    Group Size  Group Distance
   1.00000000000000                      15   0.213333333333333E-001
   2.00000000000000                      15   1.33561904761905

Results
   Delta Observed  = 0.678476190476191
   Delta Expected  = 0.664275862068965
   Delta Variance  = 0.255788784003516E-003
   Delta Skewness  = -0.989342490484899

              Standardized test statistic = 0.887886882113226
                 Probability (Pearson Type III) of a
                    smaller or equal delta = 0.814363486267441
```

Now the *P*-value is quite large (0.81) indicating that it is likely that these samples come from the same population, i.e., there is no difference between the groups. The variances of the 2 groups differ considerably as evidenced by the average within group distance (when squared Euclidean distances are used this value is twice the variance). The medians are still 15.10 and 15.40, respectively, but the means now are 15.09 and 15.23, respectively. The parametric two-sample *t*-test also results in a large *P*-value (0.54). The reason for the discrepancy in results for data in which only one value is changed is the use of squared distance. In the squared Euclidean distance analysis space the distance of the outlier from the bulk of the data is exaggerated because it is squared. Now compare the results of analyzing the data of Example 4 in a space corresponding to the geometric space of the data itself. Issue the following command after using the data in EXAMPLE4.DAT.

>MRPP RESPONSE * GROUP / V=1 C=1

which, since these are the default values, is equivalent to

>MRPP RESPONSE * GROUP

Here are the results (EXAMPLE4B.OUT).

```
              Multi-Response Permutation Procedure (MRPP)

Data Used
                   Data File: EXAMPLE4.DAT
       Grouping Variable: GROUP
      Response Variables: RESPONSE

Specification of Analysis
   Number of observations: 30
           Number of groups: 2
          Distance exponent: 1.00000000000000
```

```
        Weighting factor: n(I)/sum(n(I)) = C(I) = 1

Group Summary
   Group Value                    Group Size   Group Distance
   1.00000000000000                      15    0.116190476190476
   2.00000000000000                      15    0.531428571428572

Results
   Delta Observed  = 0.323809523809524
   Delta Expected  = 0.418390804597701
   Delta Variance  = 0.600024745882859E-004
   Delta Skewness  = -2.36855793079810

              Standardized test statistic = -12.2101390555564
                Probability (Pearson Type III) of a
                  smaller or equal delta = 0.626210563713154E-005
```

Now the resulting P-value (0.0000063) is in line with the results obtained from the data without the single aberrant value. This is a demonstration of the sensitivity of variance (squared Euclidean distance) based statistics and estimates of means to even a single outlying value. Estimates of medians and statistics based on absolute deviations (Euclidean distance) are far less sensitive to outlying data observations (Mielke and Berry 2001).

Here is another example of how it is possible to get varying statistical results by methods that differ in their underlying geometry. The distance and elevation change (in meters) for male and female blue grouse (*Dendragapus obscurus*) migrating from where they were marked on their breeding range to their winter range are given in the data file BGROUSE.DAT and are plotted in Figure 5 (data from Cade and Hoffman 1993). Generally the males seem to migrate farther and higher than the females and distance moved and elevation change are correlated ($r = 0.71$).

To test gender differences in both distance and elevation, the multivariate parametric test is Hotelling's T^2, which gives $P = 0.033$ for $F = 4.145$ with $df = 2, 18$, indicating some evidence of a difference in the bivariate means (males = 13388.9, 493.0; females = 5966.7, 231.66, distance and elevation respectively). To perform a permutation version of Hotelling's T^2, you would issue the following commands:

```
>USE BGROUSE.DAT
>MRPP DIST ELEV * SEX/HOT V = 2 C = 2 EXACT
```

where the options HOT indicated Hotelling's variance/covariance standardization of the multiple dependent variables, V = 2 requests squared Euclidean distances, and C = 2 requests that groups be weighted by their relative degrees of freedom, and EXACT requests a complete enumeration of all possible permutations for computing P-values.

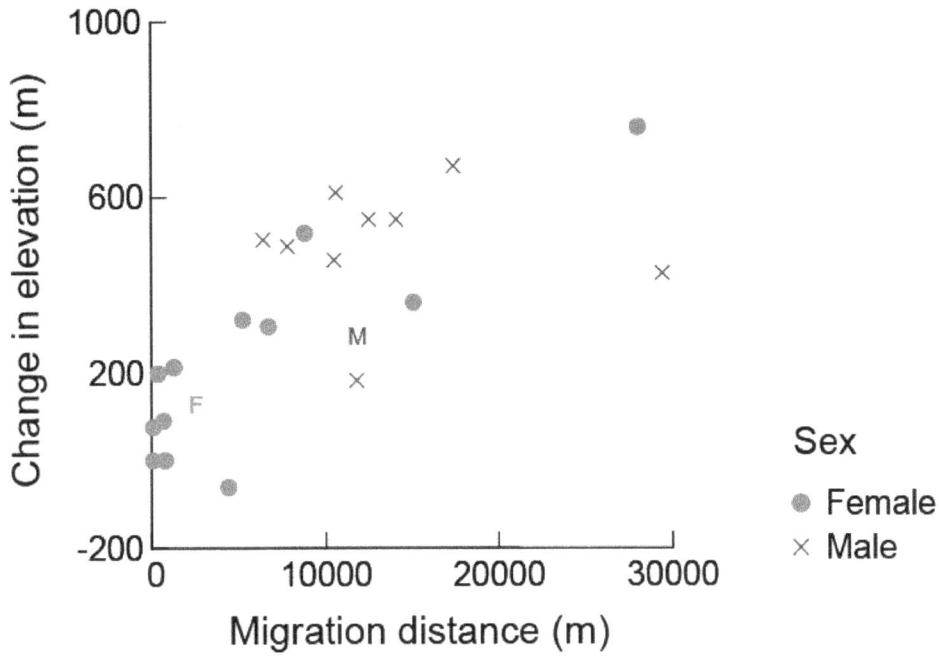

Figure 5. Migration distance and elevation change for 9 male and 12 female blue grouse (from Cade and Hoffman 1993). M and F denote bivariate medians for males and females, respectively.

Here are the results:

```
            Exact Multivariate Hotelling-type Permutation Test

Data Used
                    Data File: bgrouse.dat
   Grouping Variable: SEX
   Response Variables: DIST, ELEV

Specification of Analysis
   Number of observations: 21
          Number of groups: 2
        Distance exponent: 2.00000000000000
        Weighting factor: (n(I)-1)/sum(n(I)-1) = C(I) = 2

Group Summary
   Group Value                 Group Size
   3.00000000000000                  9
   4.00000000000000                 12

Hotelling's Commensuration Applied to Variable Values.

Results
                    Observed delta = 0.177330608239245
   Probability (Exact) of a smaller or equal delta = 0.296295036233117E-001
```

```
Variance/covariance Matrix:
      For Variables:
         Variable 1: DIST    1412312380.95238        28404633.3333333
         Variable 2: ELEV    28404633.3333333        1134020.66666667
```

Notice that there is little difference between the *P*-values for the permutation (0.030) and parametric normal theory (0.033) versions of Hotelling's T^2 for this data.

Now if we want to analyze these data in the more natural Euclidean distance space, we can issue the following commands:

>MRPP DIST ELEV * SEX/EXACT

which uses the default average Euclidean distance of each variable, ignoring the group structure, to standardize the variables so that they have an average pairwise Euclidean distance ($\Delta_{i,j}$) = 1.0. Although distances and elevation changes are in the same units (meters) so that we might consider not commensurating the variables (NOCOM option), there is some correlation between distance moved and elevation change so that it is possible that commensuration will provide more powerful hypothesis tests (Mielke and Berry 1999, 2001). Here are the results:

```
            Exact Multi-Response Permutation Procedure (EMRPP)
Data Used
            Data File: bgrouse.dat
      Grouping Variable: SEX
   Response Variables: DIST, ELEV

Specification of Analysis
   Number of observations: 21
         Number of groups: 2
         Distance exponent: 1.00000000000000
         Weighting factor: n(I)/sum(n(I)) = C(I) = 1

Group Summary
   Group Value                    Group Size
   3.00000000000000                       9
   4.00000000000000                      12

Variable Commensuration Summary
      Variable Name              Average Distance (Euclidean if V=1)
      DIST                       9264.76190476191
      ELEV                       279.228571428571

Results
                           Observed delta = 1.25745647065724
      Probability (Exact) of a smaller or equal delta = 0.316742081447964E-002
```

The same analysis but without any commensuration (NOCOM option) produced a $P = 0.008$, over twice the size of the above analysis with average Euclidean distance commensuration. Notice that the P-value with the MRPP statistic based on Euclidean distances ($V = 1$) and average Euclidean distance commensuration is an order of magnitude smaller ($P = 0.003$) than for the permutation version of Hotelling's T^2 ($P = 0.030$) based on squared Euclidean distances ($V = 2$) and the variance/covariance commensuration. There are several contributing factors. Notice, that the bivariate medians for males and females in Figure 5 indicated that the centroids of the groups were shifted in the same direction as the correlation between distance (DIST) and elevation change (ELEV). Simulations conducted by Mielke and Berry (1999) demonstrated that the average Euclidean distance commensuration of bivariate variables provided greater power than the variance/covariance standardization when the group structure was shifted parallel to the covariance structure of the 2 variables. Furthermore, since the MRPP comparisons with $V = 1$ focus on shifts in the bivariate medians which were separated by 9,271.6 m rather than shifts in the bivariate means which were only separated by 7,426.8 m, there was a larger estimated effect size for the Euclidean distance compared to the squared Euclidean distance analysis. For these data, the analysis based on Euclidean distances and bivariate medians was more powerful with greater estimated effect sizes (shift in bivariate medians). When the groups are shifted orthogonal to the covariance structure of the dependent variables, then MRPP analyses with Hotelling's variance/covariance standardization (option HOT) and $V = 1$ can be more powerful. The bivariate medians for the blue grouse movements in Figure 5 were estimated by giving the following command:

```
>MEDQ DIST ELEV*SEX/SAVE
```

where the SAVE option stores the distance between each observation and its group bivariate median (column labeled DIST2MVM) into a data file (BGROUSE.MQD) that can be used for additional analysis or graphing. The output are:

```
        2-Dimensional Median and Distance Quantiles
Data Used
    Data File: bgrouse.dat
     Grouping Variable: SEX
    # Report Variables: 2
       Report Variables: DIST, ELEV

  Specification of Analysis
    Total Number of observations: 21
                    Number of groups: 2
  -----
Results for Group Value: 3.00000000000000
   Observations in Group: 9
       Iterations to Solution: 90
          Solution Tolerance: 0.160000000000000E-010

 Within Group Median Coordinates for Variables
                Variable Name  Multivariate Median Coordinate
                     DIST  11797.1821746481
                     ELEV  292.206308872680
```

```
2-Dimensional Distance From Median Quantiles:
    Group Average Distance to Multivariate Median: 4260.34011405493
       Quantile                    Distance from Median
       0.00              [Minimum]  109.242656674530
       0.05000000000000             109.242656674530
       0.01000000000000E+01         109.242656674530
       0.25000000000000             1238.64360871698
       0.50              [Median]   2317.09148213042
       0.75000000000000             5401.29701151104
       0.90000000000000             17603.3339095665
       0.95000000000000             17603.3339095665
       1.00              [Maximum]  17603.3339095665


    - - - - -
 Results for Group Value: 4.00000000000000
    Observations in Group: 12
   Iterations to Solution: 500
       Solution Tolerance: 0.160000000000000E-010

Within Group Median Coordinates for Variables
               Variable Name  Multivariate Median Coordinate
                       DIST    2526.84016409665
                       ELEV    139.368362056321

2-Dimensional Distance From Median Quantiles:
    Group Average Distance to Multivariate Median: 5404.58960888227
       Quantile                    Distance from Median
       0.00              [Minimum]  1229.04776406248
       0.05000000000000             1229.04776406248
       0.0100000000000E+01          1732.45504780921
       0.25000000000000             1883.84586719683
       0.50              [Median]   2429.25301534112
       0.75000000000000             6284.57605921377
       0.90000000000000             12575.0954898496
       0.95000000000000             25480.7192923492
       1.00              [Maximum]  25480.7192923492


Distances to multivariate median were written to labelled file "bgrouse.MQD"
```

The bivariate median coordinates are given for the 2 variables (DIST and ELEV), and summary quantiles are provided for the distances between observations and the bivariate median for each group. The average distances to the bivariate median differ for males (4,260.3) and females (5,404.6), suggesting that there may be dispersion differences being detected by the MRPP analysis as well as shifts in bivariate medians. It is possible to test for equality of multivariate dispersions using a permutation version of a modification of Van Valen's (1978) test; the effect of the shift in group centroids removed are made with the multivariate medians rather than the multivariate means. This is accomplished for the blue grouse movements by performing a permutation version of the 2-sample t-test on the distances from the bivariate medians (variable DIST2MVM) by sex in the file saved from the previous command:

```
>USE BGROUSE.MQD
>MRPP DIST2MVM * SEX/ V = 2 C = 2 EXACT
```

The output below suggests there is little statistical support for dispersion differences.

```
              Exact Multi-Response Permutation Procedure  (EMRPP)

Data Used
                    Data File: bgrouse.MQD
   Grouping Variable: SEX
 Response Variables: DIST2MVM

Specification of Analysis
    Number of observations: 21
          Number of groups: 2
          Distance exponent: 2.00000000000000
          Weighting factor: (n(I)-1)/sum(n(I)-1) = C(I) = 2

Group Summary
   Group Value                    Group Size
   3.00000000000000                    9
   4.00000000000000                   12

Results
                        Observed delta = 82227845.9603918
   Probability (Exact) of a smaller or equal delta = 0.708165209403600
```

Note that tests for equality of univariate dispersions based on the median modification of Levene's test (Good 2000) can also be performed by requesting the univariate medians be calculated for each group with MEDQ, saving the distances from the group medians into a data file, and then comparing those distances (DIST2MVM) with the permutation version of the t-test implemented in MRPP by using the V = 2, C = 2 options. Testing for equality of dispersions after removing the effect of the estimated medians is one of those special cases where tests based on squared deviations (V = 2) have better statistical performance than using Euclidean distances (V = 1).

Because the sample size is only 21 for the blue grouse data, all the examples used the optional EXACT enumeration of all permutations to compute probabilities. This is not practical to do with larger sample sizes and by default MRPP would use the Pearson Type III moments approximation. The following command yields the default approximation:

```
>MRPP DIST ELEV * SEX
```

The output is:

```
              Multi-Response Permutation Procedure  (MRPP)

Data Used
                    Data File: BGROUSE.DAT
     Grouping Variable: SEX
   Response Variables: DIST, ELEV

Specification of Analysis
    Number of observations: 21
```

```
                 Number of groups: 2
                 Distance exponent: 1.00000000000000
                  Weighting factor: n(I)/sum(n(I)) = C(I) = 1

Group Summary
   Group Value                      Group Size  Group Distance
   3.00000000000000                          9  1.07214652525827
   4.00000000000000                         12  1.39643892970427

Variable Commensuration Summary
   Variable Name                  Average Distance (Euclidean if V=1)
   DIST                           9264.76190476191
   ELEV                           279.228571428571

Results
   Delta Observed  = 1.25745647065599
   Delta Expected  = 1.51256336315532
   Delta Variance  = 0.270618755524093E-002
   Delta Skewness  = -2.09758982733399

            Standardized test statistic = -4.90391852737653
              Probability (Pearson Type III) of a
                smaller or equal delta = 0.298316800991671E-002
```

Alternatively, we can approximate the probabilities by Monte Carlo resampling with the command:

>MRPP DIST ELEV * SEX/NPERM = 10000

where the option NPERM specifies that 9,999 random samples + the 1 observed test statistic are to be used to approximate the probabilities. The output is (BGROUSE6.OUT):

```
                Multi-Response Permutation Procedure (MRPP)
                            With Resampling

Data Used
          Data File: BGROUSE.DAT
   Grouping Variable: SEX
  Response Variables: DIST, ELEV

Specification of Analysis
   Number of observations: 21
          Number of groups: 2
          Distance exponent: 1.00000000000000
          Weighting factor: n(I)/sum(n(I)) = C(I) = 1
         Random Number Seed: 3086554
          Number of Samples: 10000

Group Summary
   Group Value                      Group Size  Group Distance
   3.00000000000000                          9  1.07214652525827
   4.00000000000000                         12  1.39643892970427

Variable Commensuration Summary
   Variable Name                  Average Distance (Euclidean if V=1)
```

```
        DIST                        9264.76190476191
        ELEV                        279.228571428571

    Results
        Delta Observed = 1.25745647065599

        Probability (Resample)of a smaller or equal delta = 0.310000000000000E-002
```

Notice that with these data that the exact, Pearson Type III approximation, and Monte Carlo resampling approximation all yield very similar *P*-values even though sample sizes were only *n* = 9 and *n* = 12.

If the data given to Blossom have been rank transformed (substituting the original values by their rank order), then MRPP can be used to emulate some well known nonparametric rank tests. Using ranks combined with the selection of V = 2 and C = 2 produces these analyses. Analyze the data from EXAMPLE4.DAT, which have been rank transformed in the file EX4RANK.DAT, with a permutation versions of the Mann-Whitney-Wilcoxon test as follows.

```
>USE EX4RANK.DAT
>MRPP RANK * GROUP /V=2 C=2


                  Multi-Response Permutation Procedure (MRPP)

    Data Used
            Data File: EX4RANK.DAT
      Grouping Variable: GROUP
    Response Variables: RANK

    Specification of Analysis
        Number of observations: 30
            Number of groups: 2
            Distance exponent: 2.00000000000000
            Weighting factor: (n(I)-1)/sum(n(I)-1) = C(I) = 2

    Group Summary
        Group Value                Group Size  Group Distance
        1.00000000000000                  15   43.2047619047619
        2.00000000000000                  15   103.133333333333

    Results
        Delta Observed = 73.1690476190476
        Delta Expected = 151.896551724138
        Delta Variance = 55.3406453148045
        Delta Skewness = -2.57249416778241

                Standardized test statistic = -10.5828922341408
                    Probability (Pearson Type III) of a
                        smaller or equal delta = 0.400568453547526E-004
```

If there are more than three groups the test is analogous to the Kruskal-Wallis one-way analysis of variance by ranks. Note that both these tests are for univariate data (one response variable), but MRPP also is able to analyze multivariate data (ranked or unranked) as well, offering a generalization of these tests. Further, the approximation used by MRPP is more accurate than the normal approximation used by the classical rank tests, since it uses the skewness of the probability distribution in the Pearson Type III approximation. Of course, it is also possible to approximate the probabilities with the Monte Carlo resampling option. Since these tests use $V = 2$ and $C = 2$, they are not congruent with the data space. Use the default values of V and C to produce a congruent analysis. Thus besides generalizing some standard nonparametric tests to multiple dependent variables, MRPP adds congruent Euclidean distance variants to the statistical repertoire.

The TRUNC = # (truncation) option, if given on the MRPP command line, causes the MRPP analysis to replace interobject distances ($\Delta_{i,j}$) greater than the truncation value (call it B) with the truncation value ($\Delta_{i,j} = \Delta_{i,j} : \Delta_{i,j} < B; \Delta_{i,j} = B : \Delta_{i,j} \geq B$). For example,

```
>MRPP VAR1 VAR2 * GROUP / TRUNC = 55
```

will replace distances greater than 55 with 55 in the permutation calculations. This is useful for detecting pattern and group clustering where one (or more) of the groups itself clusters in more than one region of the analysis space and another group is distributed uniformly or randomly in the same space. The truncation value (e.g., 55) specified is the average diameter of the sub-clusters. Data plotting and experimentation with truncation values are advised. Examples where truncation is useful include: One kind of archeological artifact may be found in two distinct areas of a site while another artifact type is found scattered throughout the site. Clumping of plants in a homogeneous site or pattern of habitat types within a landscape are detectable with a truncated MRPP analysis (Reich et al. 1991). For further information see Mielke (1991).

The EXCESS option allows for several comparisons not possible with other statistical procedures. MRPP takes data that, before analysis, are classified into groups. In the usual case the groups represent comparable levels of classification (e.g., male-female; treatments a, b, and c; or before and after observations). But in some cases one of the groups may not be comparable to the other groups of interest. This happens for example when one group is considered miscellaneous or otherwise contains unclassifiable objects. When such a group exists it may, in MRPP, be treated as an excess group. Since the concept of an excess group is not dealt with by most familiar statistical methods, a few examples will help clarify the idea.

In a study of the spatial distribution of artifacts in an archeological site Berry et al. (1983) note that many times artifacts can not readily be classified. A particular artifact may be anomalous, lack sufficient defining characteristics, or be broken or too worn to be classifiable. Such objects are definitely artifacts and may contain information, yet treating such a class on equal footing with other well defined artifact classes seems inappropriate. Investigators usually have the choice of excluding such miscellaneous classes from analysis or including them and risking bias in results or interpretation. MRPP gives the additional choice of including the excess group, but without elevating its status to that of the other groups. The observations of the excess group are treated as background noise, against which the observations on the other groups are analyzed.

Another example of the use of an excess group concerns the presence of higher lead concentration in soils near the center of a city (Mielke et al. 1983). The locations (x and y spatial coordinates) of high concentration soil samples ($\geq$ median) were compared with the locations of all samples, low and high concentration, to determine whether higher concentrations of lead are associated with the city center.

In the excess group MRPP with a group of size n and an excess group of size m an intragroup average distance is computed for each possible combination of n observations out of the $n + m$ possible observations. These values comprise the distribution of the test statistic, delta, to which is compared the actual intragroup distance.

The excess group can be implemented in comparisons of used versus available resources for a particular organism in a design where a random sample of resources is obtained and then presence (used) and absence (unused) observed. The used habitats are alike in that they all share the features necessary for the organism's survival. But the unused habitats may not form such a unitary group, some may be suitable for the organism and just happen not to be used, others may not be suitable at all, and among these some may not be suitable for lack of one requirement and others for lack of another requirement.

Here is an example comparing used versus available blue grouse habitat described by the basal area measurements of four kinds of trees present in stands on winter range (data from Cade and Hoffman 1990). Note that the $n = 16$ forest stands measured are an exhaustive and exclusive partitioning of the finite population of habitats studied (i.e. no random sampling assumptions apply).

```
>USE HABITAT.DAT
>TITLE Basal Area of Douglas Fir, Juniper, Aspen, and Other
>MRPP DFIR JUNIP ASPEN OTHER * USE / EXCESS NOCOM
```

Here are the results:

```
              Multi-Response Permutation Procedure (MRPP)

 Data Used
          Data File: HABITAT.DAT
   Grouping Variable: USE
  Response Variables: DFIR, JUNIP, ASPEN, OTHER

 Specification of Analysis
    Number of observations: 16
          Number of groups: 1
          Distance exponent: 1.00000000000000
          Weighting factor: n(I)/sum(n(I)) = C(I) = 1

 Group Summary
     Group Value                    Group Size   Group Distance
     1.00000000000000                      12    9.16824455483135
     2.00000000000000*                      4*
```

```
    * Excess group

Variables are not commensurated

Results
   Delta Observed = 9.16824455483135
   Delta Expected = 10.0781647461370
   Delta Variance = 1.26844531569669
   Delta Skewness = -0.531303217124851

              Standardized test statistic = -0.807918267201095
                 Probability (Pearson Type III) of a
                 smaller or equal delta = 0.199433916249275
```

In this example the used habitats do not seem to differ ($P = 0.200$) in tree basal area from the available (i.e., used plus unused) habitats. NOCOM was selected for no variable commensuration because tree basal areas were all in the same units (square meters/ha) and occurred at the same scale (tens of square meters/ha). However, there is some covariation among the basal areas, so commensurating them with the average Euclidean distance may be desirable. Use of the average Euclidean distance here leads to even less difference with an exact $P = 0.896$.

The ARC = *num* option allows an analysis to be conducted on univariate circular data such as time or compass orientation. This analysis recognizes that there are no endpoints to the measurement scale. Distances between replicates used in the ARC analyses are the shorter of the 2 possible distances around the circular distribution, i.e. min ($|x_i - x_j|$ and ARC - $|x_i - x_j|$). The ARC = *num* specifies the number of units in the circular distribution so that input data can be standardized to values on a unit circle. The ARC = *num* command submits the standardized data to an MRPP program configured for circular distributions.

As an example, consider an analysis of the orientation of movements of striped newts (*Notophtalmus peristriatus*) immigrating to and emigrating from Breezeway Pond, Florida in 1985 - 1990 (Dodd and Cade 1998). Figure 6 presents the angular orientation of 585 females immigrating to and 564 emigrating from the pond that were captured in pitfall buckets inside and outside of a drift fence surrounding the pond.

Select the data file and implement the arc-distance analysis with the following commands.

```
>USE NPOF.DAT
>MRPP ANGLE * EI/ ARC=360
```

The grouping variable EI has 1's for emigrating and 2's for immigrating females. Here are the results of this analysis:

```
              Multi-Response Permutation Procedure (MRPP)

Data Used
          Data File: NPOF.DAT
   Grouping Variable: EI
  Response Variables: ANGLE
```

```
Specification of Analysis
   Number of observations: 1149
          Number of groups: 2
         Distance exponent: 1.00000000000000
          Weighting factor: n(I)/sum(n(I)) = C(I) = 1
        ARC distances used: 360.000000000000
                            Intervals in unit circle

Group Summary
   Group Value                      Group Size  Group Distance
   1.00000000000000                        585  89.1737501463529
   2.00000000000000                        564  89.5317889220614

Results
   Delta Observed = 89.3494976393900
   Delta Expected = 89.7432610693134
   Delta Variance = 0.417749337172887E-002
   Delta Skewness = -1.96939238666513

             Standardized test statistic = -6.09224688551458
                 Probability (Pearson Type III) of a
                 smaller or equal delta = 0.796576837592333E-003
```

The ARC analyses indicated that immigration and emigration orientation of the striped newts differed ($P = 0.008$). More females immigrated to the northeast and southwest, whereas more emigrated from the southeast and northwest. The arc-distance analyses with MRPP are likely to be better than the more conventional Watson's test, especially useful when comparing circular distributions that have unequal angular variation or that are multimodal (Mielke and Berry 2001). The ARC option in Blossom is intended to be used with any univariate cyclical data (angular orientiation, days of the year, hour of the day); more complicated transformations are possible for spherical data and combinations of scalar and circular data (see Mielke 1986, and Mielke and Berry 2001).

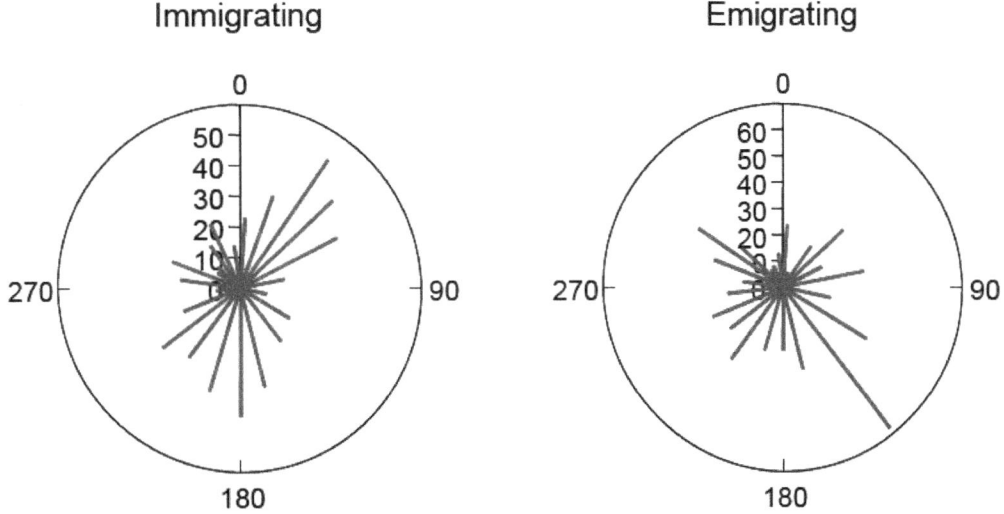

Figure 6. Pattern of immigration and emigration for female striped newts at Breezeway Pond, Florida, 1985-1990. Length of the lines indicate number of newts counted in pitfalls (data from Dodd and Cade 1997).

Multiresponse Randomized Block Procedure (MRBP)

Data from a complete randomized block design or data that can be construed in a treatment by block manner can be analyzed by specifying a blocking variable on the MRPP command line. The following data (Mielke and Iyer 1982) are from a mine reclamation study comparing oven-dried biomass (gm) of 3 species of shrubs in 6 treatments (1 = no fertilizer, 2 = low fertilizer, 3 = high fertilizer, 4 = mulch and no fertilizer, 5 = mulch and low fertilizer, and 6 = mulch and high fertilizer) by 3 blocks (different plots). A complete randomized block analysis is done with the following commands:

```
>USE MRBP.DAT
>MRPP SPP1 SPP2 SPP3 * TRTMT * BLOCK
```

Here are the results of the MRBP analysis with the default multivariable commensuration and block alignment. Note, the original analysis by Mielke and Iyer (1982) did not commensurate or align the data and you can duplicate their analysis by using the options /NOALIGN NOCOM.

```
Multi-Response Permutation Procedure for Blocked Data (MRBP)

Data Used
                    Data file: MRBP.DAT
      Grouping Variable: TRTMT
       Blocking Variable: BLOCK
   Response Variables: SPP1, SPP2, SPP3

Specification of Analysis
   Number of observations: 18
          Number of groups: 6
          Number of blocks: 3
          Distance exponent: 1.00000000000000

Group Summary
   Group Value                     Group Size
   1.00000000000000                    3
   2.00000000000000                    3
   3.00000000000000                    3
   4.00000000000000                    3
   5.00000000000000                    3
   6.00000000000000                    3

Block Alignment Summary
   Block Value                 Variable Name          Alignment Value
   1.00000000000000            SPP1                   6.50000000000000
                               SPP2                   3.16500000000000
                               SPP3                   2.17000000000000
   2.00000000000000            SPP1                   9.91500000000000
                               SPP2                   1.16500000000000
                               SPP3                   2.66500000000000
3.00000000000000               SPP1                6.25000000000000
                               SPP2                   1.91500000000000
                               SPP3                   2.41500000000000

Variable Commensuration Summary
   Variable Name                 Average Euclidean Distance
```

```
   SPP1                        7.60150326797386
   SPP2                        3.10692810457516
   SPP3                        0.900588235294119

Results
   Delta Observed = 1.78519097155486
   Delta Expected = 1.98049119623354
   Delta Variance = 0.209317316371645E-001
   Delta Skewness = -0.389741935641221

             Agreement measure among blocks = 0.986120135500246E-001
                  Standardized test statistic = -1.34989554442147
                     Probability (Pearson Type III) of a
                        smaller or equal delta = 0.949929802101351E-001
```

The *P*-value is 0.095, indicating weak evidence to reject the null hypothesis of no treatment effect. The original analysis without commensurating and aligning variables gave $P = 0.067$. Because of the small number of blocks and treatments it is possible to conduct this analysis by complete enumeration of the permutation distribution by using the option EXACT. This yields $P = 0.099$. The Monte Carlo resampling approximation also is available for problems with large block and treatment structure.

The data used in the MRBP test have been aligned so that the median of the blocks are all equal. The value chosen to align each block is selected to make the block medians all equal to zero. If there is more than one response variable then Blossom adjusts or commensurates variables by their average Euclidean distance by default as in MRPP. The block alignment values and variable commensuration values are reported.

It is possible to turn off one or both of the alignment and variable commensuration options. The NOALIGN option given anywhere after the slash (/) of the MRPP command produces an analysis without data alignment. The NOCOM option given anywhere after the slash produces an analysis without multivariate commensuration. These options can be important for special applications of MRBP. Here is an example command line:

>MRPP LENGTH * GROUP * BLOCK / NOALIGN

Of course since only 1 variable, LENGTH, was specified, no variable commensuration is done. This option is especially useful when the blocked design is used not so much to detect treatment effects but to get a measure of the agreement among blocks. One use for this option is numerical model verification. Here blocks contain the predictions of one or more models and one block contains measured results. See Tucker et al. (1989) for details. Agreement measures (1 - observed delta/expected delta) based on Euclidean distances are generalizations of Cohen's kappa extended to multiple groups, multiple variables, and interval data (Berry and Mielke 1988). The agreement measure based on squared Euclidean distances ($V = 2$) applied to interval data is a linear transform of Pearson's correlation coefficient, i.e., a probability value for a correlation coefficient based on a permutation argument can be obtained.

Here is an example analysis comparing measures of the proportion of basal area to the proportion of canopy cover of lodgepole pine (*Pinus contorta*) in 31 stands of subalpine forest in Colorado(Fig. 7) (Cade 1997).

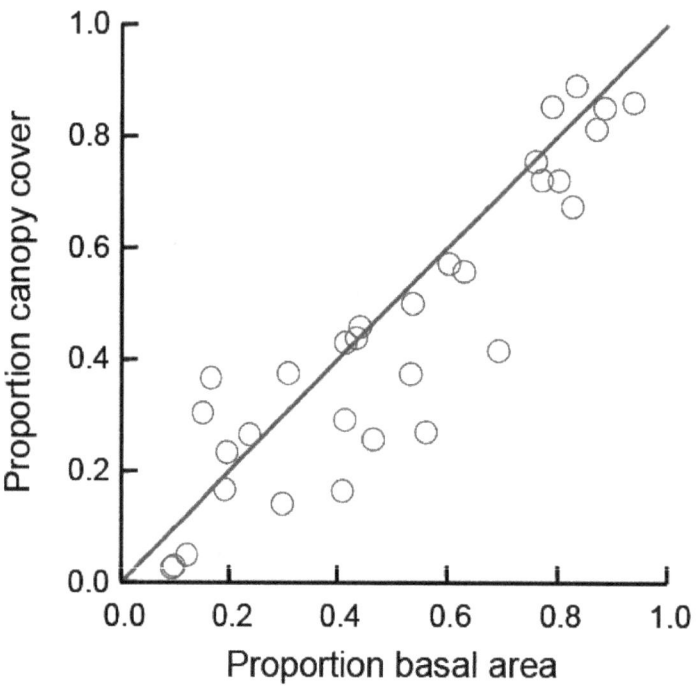

Figure 7. Proportion of basal area and canopy cover for lodgepole pine in 31 stands (data from Cade 1997). Solid line corresponds to perfect agreement.

The 31 samples plots are specified by the grouping variable STAND and the proportion of either basal area or canopy cover are specified by the blocking variable METHOD. PCTLCC is the response variable for proportion lodgepole pine.

```
>USE AGREE2.DAT
>MRPP PCTLCC * STAND * METHOD/ NOALIGN
```

Here are the results of the analysis:

```
Multi-Response Permutation Procedure for Blocked Data (MRBP)

Data Used
            Data file: AGREE2.DAT
  Grouping Variable: STAND
  Blocking Variable: METHOD
Response Variables: PCTLCC
```

```
Specification of Analysis
   Number of observations: 62
         Number of groups: 31
         Number of blocks: 2
         Distance exponent: 1.00000000000000

Group Summary
    Group Value                        Group Size
    1.00000000000000                        2
    2.00000000000000                        2
    3.00000000000000                        2
    4.00000000000000                        2
    5.00000000000000                        2
    6.00000000000000                        2
    7.00000000000000                        2
    8.00000000000000                        2
    9.00000000000000                        2
   10.0000000000000                        2
   11.0000000000000                        2
   12.0000000000000                        2
   13.0000000000000                        2
   14.0000000000000                        2
   15.0000000000000                        2
   16.0000000000000                        2
   17.0000000000000                        2
   18.0000000000000                        2
   19.0000000000000                        2
   20.0000000000000                        2
   21.0000000000000                        2
   22.0000000000000                        2
   23.0000000000000                        2
   24.0000000000000                        2
   25.0000000000000                        2
   26.0000000000000                        2
   27.0000000000000                        2
   28.0000000000000                        2
   29.0000000000000                        2
   30.0000000000000                        2
   31.0000000000000                        2

Data are not aligned within blocks

Results
    Delta Observed = 0.943115334584194E-001
    Delta Expected = 0.306180938705266
    Delta Variance = 0.121939095361522E-002
    Delta Skewness = -0.826612374970986E-001

          Agreement measure among blocks = 0.691974510701972
               Standardized test statistic = -6.06731807413269
                    Probability (Pearson Type III) of a
                    smaller or equal delta = 0.865316549394205E-008
```

The agreement measure in this analysis (0.692) indicates that there is an average reduction in Euclidean distance between the proportions of basal area and canopy cover that is 69% greater than expected by chance and this differs from zero with $P < 0.0001$. The observed delta = 0.094 which indicates that the 2 proportionate measures of lodgepole pine differed on average by 0.094 across all 31 stands (Fig. 7). There was good but not perfect agreement between measures of the proportion of basal area and the proportion of canopy cover for characterizing the lodgepole pine contribution to the forest composition. Additional univariate agreement comparisons for subalpine fir (*Abies lasiocarpa*) and Engelmann spruce (*Picea engelmannii*) are given in Cade (1997). A multivariate measure of agreement that considers all 3 species simultaneously given in Cade (1997) is performed with the command:

MRPP PCTSCC PCTFCC PCTLCC * STAND * METHOD/ NOCOM NOALIGN

The results indicate that the average deviation between proportionate measures of basal area and canopy cover is 0.168 (observed delta) across the 31 stands for the 3 conifer species and the agreement measure indicates a 62% reduction in the observed deviation over that expected by chance.

```
Multi-Response Permutation Procedure for Blocked Data (MRBP)

Data Used
          Data file: AGREE2.DAT
 Grouping Variable: STAND
 Blocking Variable: METHOD
Response Variables: PCTSCC, PCTFCC, PCTLCC

Specification of Analysis
   Number of observations: 62
          Number of groups: 31
          Number of blocks: 2
          Distance exponent: 1.00000000000000

Group Summary
   Group Value                    Group Size
      1.00000000000000                    2
      2.00000000000000                    2
      3.00000000000000                    2
      4.00000000000000                    2
      5.00000000000000                    2
      6.00000000000000                    2
      7.00000000000000                    2
      8.00000000000000                    2
      9.00000000000000                    2
     10.0000000000000                     2
     11.0000000000000                     2
     12.0000000000000                     2
     13.0000000000000                     2
     14.0000000000000                     2
     15.0000000000000                     2
     16.0000000000000                     2
     17.0000000000000                     2
     18.0000000000000                     2
     19.0000000000000                     2
     20.0000000000000                     2
     21.0000000000000                     2
     22.0000000000000                     2
     23.0000000000000                     2
     24.0000000000000                     2
     25.0000000000000                     2
     26.0000000000000                     2
```

```
27.0000000000000                        2
28.0000000000000                        2
29.0000000000000                        2
30.0000000000000                        2
31.0000000000000                        2
```

Data are not aligned within blocks

Variables are not commensurated

Results
   Delta Observed = 0.168138081382156
   Delta Expected = 0.440873737189001
   Delta Variance = 0.155287201162739E-002
   Delta Skewness = -0.876180008760865E-001

       Agreement measure among blocks = 0.618625317864930
         Standardized test statistic = -6.92108347758165
           Probability (Pearson Type III) of a
             smaller or equal delta = 0.116138823237641E-009

For information on other ways to align data useful for analyzing incomplete block and Latin square designs with MRBP see Fawcett (1990), Mielke and Iyer (1982), and Hodges and Lehmann (1962).

If V = 2 is chosen, then the univariate version of this test is a permutation version of analysis of variance for complete randomized blocks. Note that when V = 2 is used in an MRBP analysis that the blocks are self-aligning to a common mean and no alignment is required; analyses made with MRBP and V = 2 and the option NOALIGN should result in identical test statistics and *P*-values as when alignment is not turned off. Specification of the C (group averaging method) parameter has no effect, since group sizes have to be the same. Also the EXCESS option is not supported for MRBP and is ignored. The EXACT option is available only for some small block (<10) and group combinations. The Monte Carlo resampling approximation of *P*-values is available with the option /NPERM = *num*.

If ranked data are used and V = 2 is specified, then the test (with one response variable) is functionally related to Friedman's nonparametric randomized block analysis.

Permutation Tests for Matched Pairs (PTMP)

Matched pair tests can be performed by the MRPP command. Essentially the matched pairs test is a special case of the randomized block version of MRPP with one or more response variables, two groups, and a blocking variable identifying pairs. Data of this sort can be analyzed by an MRPP command specified just like that for performing an MRBP. For example the sample data file PAIRED1.DAT contains one response (RESPONSE), for two groups (GROUP), and with the paired members of each group indicated by a blocking variable (PAIR). Use this file and perform a matched pairs test by issuing the following command:

   >MRPP RESPONSE * GROUP * PAIR

Here are the results (PAIRED1.OUT):

```
Multi-Response Permutation Procedure for Blocked Data (MRBP)

 Data Used
            Data file: Paired1.dat
    Grouping Variable: GROUP
    Blocking Variable: PAIR
  Response Variables: RESPONSE

 Specification of Analysis
    Number of observations: 20
           Number of groups: 2
           Number of blocks: 10
          Distance exponent: 1.00000000000000

 Group Summary
    Group Value                    Group Size
    1.00000000000000                      10
    2.00000000000000                      10

 Block Alignment Summary
    Block Value               Variable Name          Alignment Value
    1.00000000000000          RESPONSE               4.27500000000000
    2.00000000000000          RESPONSE               3.34000000000000
    3.00000000000000          RESPONSE               6.54500000000000
    4.00000000000000          RESPONSE               3.07000000000000
    5.00000000000000          RESPONSE               2.88000000000000
    6.00000000000000          RESPONSE               8.19000000000000
    7.00000000000000          RESPONSE               6.10500000000000
    8.00000000000000          RESPONSE               5.06500000000000
    9.00000000000000          RESPONSE               2.69500000000000
   10.0000000000000          RESPONSE               0.79000000000000

 Results
    Delta Observed = 1.21144444444444
    Delta Expected = 1.88722222222222
    Delta Variance = 0.209924691358025E-001
    Delta Skewness = -1.98423908598235

           Agreement measure among blocks = 0.358080659405358
                Standardized test statistic = -4.66414608785788
                   Probability (Pearson Type III) of a
                   smaller or equal delta = 0.342067401604490E-002
```

With one response variable and $V = 2$ specified on the command line, then this test mimics the *t*-test for matched pairs.

Sometimes it is convenient to structure paired data such that the values for each pair are given on a single line in the data file with a separately named variable for the response of the first and of the second members of each pair. Blossom allows for this different data structure. Use the example file PAIRED2.DAT and simply issue the following command:

```
>MRPP FIRST SECOND /PAIRED
```

The PAIRED option signifies that the observations are paired (next to each other) in the data file. Thus the pairing is indicated by position not by a blocking variable. Also no grouping variable is specified because in PTMP there can only be two groups. The univariate observations for each group correspond to the columns named FIRST and SECOND. Note, this is a special data file format useful only for PTMP, which is a univariate, two group, paired comparison, where the number of blocks equals the number of pairs.

Here are the results of the above command:

```
 Multi-Response Permutation Procedure for Paired Data (PTMP)

Data Used
                              Data file: PAIRED2.DAT
    Response Variables (Treatment Groups): FIRST, SECOND

Specification of Analysis
    Number of observed pairs (Blocks): 10
                    Distance exponent: 1.00000000000000

Results
   Delta Observed = 2.42288888889131
   Number of non-zero differences = 10
   Probability (Exact) of a smaller or equal Delta = 0.390625000000000E-002
```

Because the number of pairs in this data set is less than 20 the *P*-value reported was obtained by exact enumeration of the permutation distribution (and thus differs slightly from the *P*-value given in the previous example). With more than 20 pairs an approximation with the Pearson Type III distribution is used by default or the Monte Carlo resampling option can be invoked with the option /NPERM = *num*. Notice that the different test statistic structures produce an observed delta in PTMP that is exactly twice the observed test statistic for the same problem in MRBP. Also, data in PTMP are aligned to a median of 0 by the structure of the test statistic.

It is possible to do a 1-sample comparison of data with an hypothesized parameter for central tendency (either median or mean) with PTMP by making one of the column variables equal to the hypothesized parameter and the other the observed data vector (Mielke and Berry 2001). If the hypothesized parameter is a median and PTMP is implemented with V = 1 then this test is for a null hypothesis that the sample comes from a population with median equal to the specified value. If the hypothesized parameter is a mean and PTMP is implemented with V = 2 then this test if for a null hypothesis that the sample comes from a population with mean equal to the specified value.

Mulitvariate extensions of the 1-sample comparison are made by using MRBP and specifying the vector of hypothesized parameters for the multivariate median (mean) as one group, the observed vector as the second group, for each of *n* blocks comprising the sample. As an example, consider the data on ring-necked pheasant (*Phasianus colchicus*) habitat selection from Aebischer et al. (1993: Appendix 1), where the proportion of home ranges in 5 habitat types (scrub, broadleaf woodlands, conifer woodlands, grasslands, and crops) for 13 radio-marked birds were compared to the available proportions of these habitat types. Because these data are compositions with a unit sum constraint, Aebischer et al. (1993) chose to analyze these data with log ratios in a MANOVA. We can perform a similar 1-sample analysis comparing the observed proportions of the habitat types for the 13 birds with the hypothesized available proportions in MRBP without resorting to log ratios (which are problematic when you have some zero proportions). The data file PREFER.DAT has 13 blocks (BIRD) for the grouping USE = 1 corresponding to the observations for the 13 birds, and the same 13 block values for the grouping USE = 2 corresponding to 13 replications of the hypothesized available proportions of the habitat types. Issue the commands:

>USE PREFER.DAT
>MRPP SCRUB BROAD CONIFER GRASS CROP*USE*BIRD

The following output indicated that the 13 pheasants were not using habitat types in proportion to their availability. Of course, it is possible to do a permutation version of the 1-sample MANOVA analysis on log ratios as done by Aebischer et al. (1993), but the Euclidean distance statistics of MRPP avoid concerns about singular matrices with dependent variables having the unit sum constraint and ad hoc procedures needed to deal with zero proportions when transforming to log ratios.

```
Multi-Response Permutation Procedure for Blocked Data (MRBP)

Data Used
          Data file: PREFER.DAT
 Grouping Variable: USE
 Blocking Variable: BIRD
Response Variables: SCRUB, BROAD, CONIFER, GRASS, CROP

Specification of Analysis
   Number of observations: 26
         Number of groups: 2
         Number of blocks: 13
         Distance exponent: 1.00000000000000

Group Summary
   Group Value                  Group Size
   0.00000000000000                  13
   1.00000000000000                  13

Block Alignment Summary
   Block Value                  Variable Name        Alignment Value
   1.00000000000000             SCRUB                11.4100000000000
                                BROAD                5.60000000000000
                                CONIFER              0.36500000000000
                                GRASS                26.4150000000000
                                CROP                 56.1900000000000
   2.00000000000000             SCRUB                11.9000000000000
                                BROAD                11.9650000000000
                                CONIFER              0.36500000000000
                                GRASS                26.6150000000000
                                CROP                 49.1450000000000
   3.00000000000000             SCRUB                5.77000000000000
                                BROAD                7.48000000000000
                                CONIFER              0.36500000000000
                                GRASS                55.8650000000000
```

	CROP	30.5050000000000
4.00000000000000	SCRUB	6.00000000000000
	BROAD	16.5450000000000
	CONIFER	0.3650000000000
	GRASS	32.5350000000000
	CROP	44.5400000000000
5.00000000000000	SCRUB	3.81500000000000
	BROAD	19.7600000000000
	CONIFER	5.52500000000000
	GRASS	53.9050000000000
	CROP	16.9900000000000
6.00000000000000	SCRUB	4.32500000000000
	BROAD	19.8750000000000
	CONIFER	5.42000000000000
	GRASS	53.3850000000000
	CROP	16.9900000000000
7.00000000000000	SCRUB	3.78000000000000
	BROAD	20.2350000000000
	CONIFER	5.87500000000000
	GRASS	53.1100000000000
	CROP	16.9900000000000
8.00000000000000	SCRUB	5.94000000000000
	BROAD	23.9700000000000
	CONIFER	0.3650000000000
	GRASS	52.7200000000000
	CROP	16.9900000000000
9.00000000000000	SCRUB	6.43000000000000
	BROAD	31.1950000000000
	CONIFER	0.3650000000000
	GRASS	45.0000000000000
	CROP	16.9900000000000
10.0000000000000	SCRUB	7.47000000000000
	BROAD	9.02500000000000
	CONIFER	0.3650000000000
	GRASS	66.1350000000000
	CROP	16.9900000000000
11.0000000000000	SCRUB	8.79000000000000
	BROAD	20.8950000000000
	CONIFER	0.3650000000000
	GRASS	52.9400000000000
	CROP	16.9900000000000
12.0000000000000	SCRUB	6.46000000000000
	BROAD	10.0900000000000
	CONIFER	0.3650000000000
	GRASS	66.0800000000000
	CROP	16.9900000000000
13.0000000000000	SCRUB	4.37500000000000
	BROAD	14.6550000000000
	CONIFER	2.42000000000000
	GRASS	61.5550000000000
	CROP	16.9900000000000

Variable Commensuration Summary

Variable Name	Average Euclidean Distance
SCRUB	4.99252307692307
BROAD	11.6876307692308
CONIFER	2.45584615384616
GRASS	15.0708307692308
CROP	18.2146461538461

Results
    Delta Observed = 2.23830657082488

```
Delta Expected = 2.65984307847880
Delta Variance = 0.827598480934282E-002
Delta Skewness = -1.16504990574294

      Agreement measure among blocks = 0.158481720619021
            Standardized test statistic = -4.63367266419705
             Probability (Pearson Type III) of a
                smaller or equal delta = 0.120922268783399E-002
```

We can compute the multivariate median for the proportions of the habitat types used by the 13 pheasants to compare with the hypothesized proportions by issuing the command:

>MEDQ SCRUB BROAD CONIFER GRASS CROP * USE

The output (PREFER2.OUT) indicated that the multivariate median vector for the proportions of habitats used is shifted towards a much higher proportion of broadleaf woodlands, moderately higher proportions of scrub and conifer woodlands, much lower proportions of crops, with little difference in the proportion of grasslands compared to available habitat types. Note that this summary doesn't recognize the blocked by animal nature of the design and could be made more appropriate by first taking differences between components of used and available habitat types by animal and then taking the multivariate medians of those differences.

```
         5-Dimensional Median and Distance Quantiles

Data Used
           Data File: PREFER.DAT
   Grouping Variable: USE
  # Report Variables: 5
    Report Variables: SCRUB, BROAD, CONIFER, GRASS, CROP

 Specification of Analysis
   Total Number of observations: 26
             Number of groups: 2
 -----
 Results for Group Value: 0.00000000000000
   Observations in Group: 13
  Iterations to Solution: 1
      Solution Tolerance: 0.160000000000000E-012

 Within Group Median Coordinates for Variables
                 Variable Name  Multivariate Median Coordinate
                       SCRUB      3.22000000000000
                       BROAD      9.23000000000000
                       CONIFER    0.730000000000000
                       GRASS     52.8300000000000
                       CROP      33.9800000000000

5-Dimensional Distance From Median Quantiles:
   Group Average Distance to Multivariate Median: 0.737776405560993E-014
     Quantile                Distance from Median
     0.00        [Minimum]    0.00000000000000
     0.500000000000000        0.00000000000000
     0.010000000000E+01       0.737776405560993E-014
     0.250000000000000        0.737776405560993E-014
     0.50        [Median]     0.737776405560993E-014
     0.750000000000000        0.737776405560993E-014
```

```
    0.900000000000000        0.737776405560993E-014
    0.950000000000000        0.737776405560993E-014
    1.00        [Maximum]     0.737776405560993E-014


 -----
 Results for Group Value: 1.00000000000000
   Observations in Group: 13
 Iterations to Solution: 55
      Solution Tolerance: 0.160000000000000E-012

 Within Group Median Coordinates for Variables
                 Variable Name  Multivariate Median Coordinate
                        SCRUB    7.61644708253726
                        BROAD    28.6767927313389
                        CONIFER  5.66570731938610
                        GRASS    54.0419189963013
                        CROP     3.98762906946345

 5-Dimensional Distance From Median Quantiles:
    Group Average Distance to Multivariate Median: 33.7412013141642
      Quantile                  Distance from Median
      0.00        [Minimum]     6.62127437797105
      0.500000000000000         6.62127437797105
      0.010000000000E+01        7.16431172595937
      0.250000000000000         10.4661422196434
      0.50        [Median]      30.5971449683189
      0.750000000000000         33.3721790108255
      0.900000000000000         83.1369267354442
      0.950000000000000         96.6782732275243
      1.00        [Maximum]     96.6782732275243
```

The MRPP Command Syntax

The MRPP command can take different forms depending on the nature of the analysis desired. If you don't understand MRPP consult the references given at the end of this document before attempting to change the default values. Here is the complete MRPP command syntax.

MRPP *variable list* * *grouping variable* [*(num ...)* | *(num - num)*] [**blocking variable*]

[/ V = *num* | C = *num* | EXACT | EXCESS [= *num*]| PAIRED | NOCOM | HOT|
NOALIGN |
TRUNC = *num* | ARC = *num* | NPERM [= *num*]| SEED = *num* |
SAVETEST [= *file name*]]

Items to be supplied by the user are given in lower case in italics. These are usually variable names or numbers (*num*). Items in square brackets are optional. Upper case words or letters are Blossom commands and must be entered exactly as given. The vertical line (|) can be read as "or" and separates different options that can be specified. They can be specified in any order. The optional numbers (*num*) given in parentheses after the grouping variable name specifies either a list or range of values indicating the groups to be used (these have to be numeric values of the grouping variable). If no values or range is specified, the groups correspond to each unique value of the grouping variable. To analyze blocked data (MRBP) a blocking variable is specified.

The options specified after the slash (/) control technical details of the analysis. The values for V determine the exponent of the distance function. The default is 1 and values other than V = 2 are seldom used. Valid values for C are 1, 2, 3, or 4 and determine how intragroup distances are averaged together. C = 1 is the default and corresponds to relative sample size, C = 2 corresponds to relative degrees of freedom, and the options 3 and 4 are seldom used. If the EXCESS option is specified, the excess group by default corresponds to the cases with the largest value of the grouping variable. This can be changed by adding the appropriate grouping variable value after the EXCESS option. To analyze paired data (PTMP) the PAIRED option is specified. The NOALIGN option is for blocked data analysis; the automatic alignment option is circumvented. The NOCOM option turns off default average Euclidean distance commensuration of multiple response variables. The HOT options specifies Hotelling's variance/covariance commensuration of multiple response variables. The TRUNC = *num* option is available for grouped but not blocked data. The truncation number (*num*) gives the maximum object to object distance to be used in the analysis. The ARC = *num* option provides the units of data in a circular distribution for standardization to a unit circle and inputs standardized univariate data to an arc-distance MRPP analysis. The NPERM option requests a Monte Carlo resampling approximation of *P*-values rather than the Pearson Type III moments approximation. By default NPERM uses 5,000 random samples but any number may be optionally specified by NPERM = *num*. The option SEED = *num* allows the user to specify a random number seed rather than using the default computer clock generated number. The option SAVETEST = *file name* allows you to save the Monte Carlo resampled test statistics into a column in the specified file, where the first value is always the observed test statistic.

Here are some more examples of valid MRPP command lines. They show how to perform an analysis using a subset or range of the groups indicated by the grouping variable and how to specify a particular group as the excess group. If not specified the excess group is always the group with the largest value on the grouping variable.

>MRPP W X Y Z * GROUP (2 3 6) / EXCESS

The groups used are confined to a subset of the values of the grouping variable and group 6 will be the excess group.

>MRPP VAR1 VAR2 * GP (3-8) / EXCESS = 1

The groups used are confined to a range of group values and the excess group is indicated by its value (1). The excess group value can be in the grouping variable list or range or not. For example the following three command lines produce the same analysis.

>MRPP VAR1 VAR2 * GP (3-7) / EXCESS=8

>MRPP VAR1 VAR2 * GP (3-8) / EXCESS

>MRPP VAR1 VAR2 * GP (3-8) / EXCESS=8

Terse output provided by the MRPP command following an OUTPUT /TERSE command includes the USEd file name, dependent variable names, grouping variable name, number of groups, blocking variable name (if present), observed test statistic, and *P*-value.

Multivariate Medians and Distance Quantiles (MEDQ) Command Syntax

The MEDQ command can estimate multivariate and univariate medians for grouped or ungrouped data and optionally save the distances between the observations and the medians in a file. MEDQ is intended to provided summary statistical estimates that are useful for describing group differences detected by MRPP comparisons.

MEDQ *variable list* [* *grouping variable* [*(num ...)* | *(num - num)*]]

[/ SAVE | QUANT = *num, num, ..., num*]

By default MEDQ estimates the multivariate medians (or univariate) for the variables specified in the list ignoring any group structure. If the optional group variable is specified then MEDQ computes similar estimates but by each group. Options for selecting subsets of a grouping variable work similar as in the MRPP command. If the SAVE option is specified then a file with distances between observations and estimated medians are saved to a file named with your filename and the extension .MQD. A column variable named DIST2MVM that is the distance to multivariate (or univariate) medians is stored, along with the values for variables selected, and values of any grouping variables specified for each observation. These values can be useful for graphical exploration and for conducting tests of equivalent dispersions. The option QUANT = *num, num, ..., num* allows you to specify values other than the default quantiles (min = 0.0, 0.05, 0.10, 0.25, 0.50, 0.75, 0.90, 0.95, max = 1.0) for summarizing distances to multivariate medians. Note that when you specify only a single response variable, the QUANT option also allows you to request specific univariate quantiles to be estimated as well as the default univariate median.

Terse output provided by the MEDQ command following an OUTPUT /TERSE command is identical to the default verbose output.

Multiresponse Sequence Procedure (MRSP)

The multiresponse sequence procedure command SP initiates a test of first-order serial dependency on univariate or multivariate response variables (Mielke 1991). In this analysis of ungrouped data, the agreement measure (1 – average Euclidean distance between ordered observations/average Euclidean distance among all possible pairs of observations) is a statistic describing first-order serial dependency. Significance of the null hypothesis of no first-order serial dependency is provided by the Pearson Type III approximation on the first 3 exact moments of the permutation distribution by default, optionally by exact enumeration for small samples, or by a Monte Carlo resampling procedure. In a univariate test, MRSP is analogous to the Durbin-Watson test. A permutation version of the Durbin-Watson test can be initiated by selecting the option $V = 2$ for squared Euclidean distances. In a bivariate test of animal locations, where latitude and longitude coordinates are the 2 response variables, MRSP provides a Euclidean distance analogue of Schoener's t^2/r^2 statistic (Solow 1989), which is a nonmetric measure based on squared Euclidean distances. A permutation version of Schoener's t^2/r^2 test can be implemented on bivariate data by specifying the $V = 2$ and NOCOM options. MRSP obviously provides the possibility of evaluating first-order serial dependency of >2 response variables. Blossom will commensurate (standardize) multiple variables to unit average Euclidean distance by default.

Example data of biweekly grouse locations during November through March (Cade and Hoffman 1993) are in the file BLUE162.DAT and graphed below (Fig. 8), where the numbers correspond to the temporal order of observations (variable DATE) for the response variables LAT and LONG. Multiple observations at the same location are indicated by ordered values in parentheses.

To implement an analysis of first-order serial dependency on these bivariate locational data, you issue the following commands:

```
>USE BLUE162.DAT
>SP LAT LONG * DATE/ NOCOM
```

Note that the same analysis on this data set can be implemented without specifying the sequencing variable DATE because the data set is ordered by DATE already and Blossom by default assumes the order in the data file is the sequencing variable if none is specified.

Here are the results of this analysis:

Multi-Response Sequence Permutation Procedures (MRSP)

Data Used
            Data File: Blue162.dat
        Sequenced by
    values of variable: DATE
    Response Variables: LAT, LONG

  Specification of Analysis
      Number of observations: 12
            Distance exponent: 1.00000000000000

  Variables are not commensurated

  Results
      Delta Observed = 119.655567412341
      Delta Expected = 135.466761860064
      Delta Variance = 252.038157171250
      Delta Skewness = -0.236316783882059

                  Standardized test Statistic = -0.995936231356816
                         Agreement measure = 0.116716412429317
                    Probability (Pearson Type III) of a
                      smaller or equal delta = 0.159048321416475

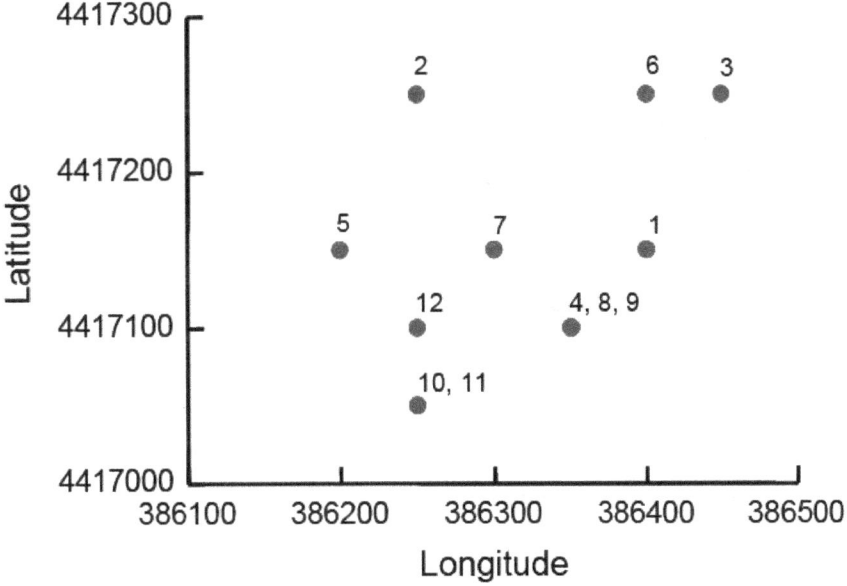

Figure 8. Latitude and longitude coordinates (m) for blue grouse (no. 162) locations on 12 dates
November – March. Numbers next to locations correspond to order of dates for locations (note
multiple dates at same locations).

The agreement measure (0.117) is interpreted as the percent reduction in average Euclidean
distance between sequentially ordered values (observed delta) over that expected without any

order (expected delta). In this analysis there is little evidence of first-order serial pattern and the null hypothesis of no serial dependency has $P = 0.159$. The expected delta is the average Euclidean distance among the locations ignoring any serial dependence (135.5 m) and the observed delta in the average Euclidean distance between sequentially ordered locations (119.7 m); both these measures are useful summary statistics describing animal home ranges (Cade and Hoffman 1993). See Mielke and Berry (2001) for descriptions of how to extend the sequence procedure to higher orders of serial dependence.

The SP Command Syntax

Here is the complete syntax for the SP command.

> SP *variable list* * [*sequencing variable*] [/ V = *num* | NOCOM | EXACT | NPERM [= *num*] | SEED = *num* | SAVETEST [= *file name*]]

The variable list is provided by the user and an optional sequencing variable can be specified in place of the grouping variable normally specified in the MRPP command syntax. If no sequencing variable is given Blossom assumes by default that the order (top to bottom) of the data in the file is the sequence desired. The sequencing variable can be a variable from the variable list. The options after the / permit you to select alternative exponents for the distance function and to turn off multivariate commensuration as in the MRPP command. Options for computing *P*-values by exact enumeration (EXACT) for small samples ($n < 10$) or by Monte Carlo resampling (NPERM | SEED) are provided just as with the MRPP command. The SAVETEST = *file name* option allows you to save the Monte Carlo resampled test statistic values into a column in the specified file, where the first value always is the observed test statistic value.

Terse output provided by the SP command following an OUTPUT /TERSE command includes the USEd file name, dependent variable names, sequencing variable name, observed test statistic (delta), agreement, and *P*-value.

Least Absolute Deviation (LAD) and Quantile Regression

LAD regression differs from least squares (OLS) regression in that the sum of the absolute, not squared, deviations of the fit from the observed values is minimized to obtain estimates. LAD regression estimates the conditional median (0.5 regression quantile) of the dependent variable (y) given independent variables (X), and its generalization, regression quantiles, estimate the conditional quantile (τ, where $0 \leq \tau \leq 1$) of y given X. Since LAD does not use squared distances, it is an obvious companion to the MRPP which emphasizes Euclidean distances. Both LAD and MRPP satisfy the congruence principle (Mielke and Berry 2001). Asymptotic distributional theory for testing procedures for LAD regression are found in Dodge (1987) and a concise, readable implementation is provided by Birkes and Dodge (1993). Cade and Noon (2003) is a primer on quantile regression for ecologists.

The LAD command is used to compute a fit of one dependent response variable by one or more independent predictor variables. The parameters in a LAD regression are tested by using a test statistic that compares the proportionate reduction in sums of absolute deviations when passing from a reduced to full parameter model (i.e., a test statistic very similar to general F-tests in OLS regression). The drop in dispersion test statistic, T_{obs}, equals (sum of absolute deviations for reduced model – sum of absolute deviations for full model) / sum of absolute deviations for full model (Cade and Richards 1996, Cade 2003, 2005). Large values of T_{obs} are evidence against the null hypothesis that the parameter(s) equal(s) zero. If all slope parameters are tested simultaneously against a reduced parameter model that includes only the intercept, then the reference permutation distribution for the test statistic T_{obs} is obtained by randomly sampling the $n!$ permutations of the dependent variable to the matrix of independent variables as described by Manly (1991) and calculating T for each permutation. However, if only a subset of parameters are being tested (partial model tests), then the reference permutation distribution for the test statistic T_{obs} is obtained by randomly sampling the $n!$ permutations of residuals from the reduced model to the matrix of independent variables and calculating T for each permutation, following (Freedman and Lane 1983). Probabilities under the null hypothesis are given by (number of $T \geq T_{obs}$ + 1)/ number of permutations sampled. Extensive power simulations demonstrated that these procedures maintained nominal error rates under the null hypothesis well across a variety of error distributions and design configurations (correlated and uncorrelated independent variables) provided the error distributions are independent and identically distributed (Cade and Richards 1996). Similar conclusions were reached for the same form of the test statistic used with OLS regression (Kennedy and Cade 1996, Anderson and Legendre 1999). The LAD permutation test is extended to any selected regression quantile (LAD is just 0.5 regression quantile) by replacing sums of absolute deviations in the test statistic computation with the appropriate sums of weighted absolute deviations used in regression quantile estimation (Cade and Richards In press).

Cade (2003, 2005) and Cade and Richards (In press) found that Type I error rates were improved when testing subsets of parameters in quantile regression models by deleting all but a single zero residual associated with the fit to $p - q$ parameters under the null hypothesis, where p is the number of parameters in the full model and q is the number of parameters being tested. As this reduces the length of the residual vector so that it no longer conforms to the $n \times p$ matrix $\mathbf{X}$ of predictors, the corresponding number of rows of $\mathbf{X}$ are randomly deleted at each permutation. This deletion of zero residuals and random deletion of rows of $\mathbf{X}$ are done by default for this drop in dispersion permutation test. In addition, Cade (2003, 2005) and Cade and Richards (In press) found that anytime the null, reduced ($p - q$) parameter model was constrained through the origin (no intercept), Type I error rates were improved by randomly recentering the residual vector since the residuals from the null model will no longer have zero associated with the specified quantile (or mean zero for OLS). This is implemented as a double permutation procedure where the first step at each iteration is to randomly recenter the selected quantile of the residual vector by a quantity generated as a random binomial for the specified quantile (e.g., 0.90). A similar operation is done for OLS regression where the quantile = 0.50 is always used to generate random binomials. The second step at each iteration then (the doubling of permutations) permutes these randomly recentered residuals to the matrix **X**. Because it is not always obvious when a model is constrained through the origin (e.g., some weighted model tests will require this and some won't), we elected to make the double permutation scheme selected by an option of the hypothesis testing command (HYP/DP).

If error distributions are not identical (heteroscedastic) then they must be transformed or weighted to be made approximately identical (homogeneous) (Cade and Richards 1996, Cade 2003, 2005, Cade and Richards In press). Cade and Noon (2003) and Cade et al. (2005a) discuss two weighting schemes, one where all quantiles have the same weights in a location\scale form of heterogeneity, and one where the weights must be estimated separately for the selected quantiles in more general models of heterogeneity. When the weights are based on a function of the independent variables (**X**), many of the permutation hypothesis tests will implicitly constrain the null model through the origin and the double permutation procedure will be required to maintain correct Type I error rates (Cade 2005, Cade et al. 2005, Cade and Richards In press).

As an alternative test for LAD and its generalization to regression quantiles, we provide a quantile rank score statistic that is less sensitive to heterogeneous error distributions (Koenker 1994, Cade et al. 1999, Koenker and Machado 1994). The permutation version of the quantile rank score test (Cade 2003, Cade et al. 2005b) maintains Type I error rates better than the asymptotic Chi-square distributional approximation (Koenker 1994) at smaller n and more extreme quantiles. It is important to note that the rank score test is not immune to the effects of heterogeneity and maintaining correct Type I error rates with this test often requires weighted estimates and test statistics just as the drop in dispersion test does (Cade 2003, Cade et al. 2005a, Cade et al. 2005b).

We will demonstrate the procedures with an example from Cade (1997), where lodgepole pine canopy cover was modeled as a function of basal area and density of the trees. Use the data file FRASERF.DAT. Issue the following command for the simple regression of canopy cover (LCC) as a linear function of basal area (APICO):

>LAD LCC = CONSTANT + APICO /TEST

The model to be computed is written out algebraically where the dependent variable is LCC (lodgepole pine canopy cover) and the single independent variable is APICO (basal area of lodgepole pine adjusted for slope of terrain). The term "CONSTANT" indicates that LAD will estimate an intercept. If "CONSTANT" is left out the fit is forced through the origin. The TEST option indicates that the model is to be compared to a reduced model that is a straight line parallel to the X axis going through the median y value (LCC). Thus, the reduced model has just one parameter, the constant. In this test Blossom uses a default sample size of 5,000 permutations (including observed value) to approximate the permutation distribution.

Here are the results of the above LAD command:

```
                Least Absolute Deviation Regression (LAD)

Data Used
     Data file: FRASERF.DAT

LAD Regression:
     LCC=CONSTANT+APICO
Results
   Number of observations: 31
       Dependent Variable: LCC

   Independent variables       Regression coefficients
   CONSTANT                        8.78874116689298
   APICO                           1.05354969353239

   Number of iterations: 3
   Sum of absolute values of the residuals: 252.851627121547
              Solution: SUCCESSFUL

Regression Evaluation:
   LAD Model:
     LCC=CONSTANT+APICO

Test Summary
     Number of permutations: 5000
         Random Number Seed: 3198580
      P-value of Full Model: 0.200000000000000E-003
```

Because canopy cover must be zero when basal area is zero, Cade (1997) used LAD regression models without an intercept term. Here the following command estimates the model above without an intercept:

LAD LCC = APICO

The output is given below:

```
                Least Absolute Deviation Regression (LAD)

Data Used
   Data file: FRASERF.DAT

LAD Regression:
     LCC=APICO
Results
   Number of observations: 31
       Dependent Variable: LCC

   Independent variables       Regression coefficients
   APICO                           1.3143404040219

   Number of iterations: 1
   Sum of absolute values of the residuals: 267.446922457515
              Solution: SUCCESSFUL
```

A multiple independent variable LAD regression is specified by adding the appropriate independent variable names to the LAD command. Here we consider the model used by Cade (1997) with lodgepole pine density (PICOPHA) as an additional explanatory variable:

>LAD LCC = APICO + PICOPHA /SAVE

The added variables are assumed to be in the data file in USE. The SAVE option causes Blossom to save a labeled data file that includes the variables in the model and two new columns that contain the predicted *y* values (PRED) and the residuals (RESID). The saved file by default has the name of the file in use but with a ".LAD" file extension. To specify the saved file's name follow the save option with a file name e.g., SAVE = MODEL1.OUT. If the save file already exists you will be prompted with a choice to overwrite it or not. If a LAD command with the SAVE option appears in a SUBMIT file any preexisting save file is automatically overwritten. Here are the results:

```
                  Least Absolute Deviation Regression (LAD)

 Data Used
    Data file: FRASERF.DAT

 LAD Regression:
     LCC=APICO+PICOPHA
 Results
    Number of observations: 31
        Dependent Variable: LCC

    Independent variables              Regression coefficients
    APICO                              0.934738765025370
    PICOPHA                            0.115723762501552E-001

    Number of iterations: 3
    Sum of absolute values of the residuals: 127.511282250334
                Solution: SUCCESSFUL
 Model, predicted, and residual values saved in labelled file: FRASERF.LAD
```

The regression function, observed values and residuals are plotted in Fig. 9.

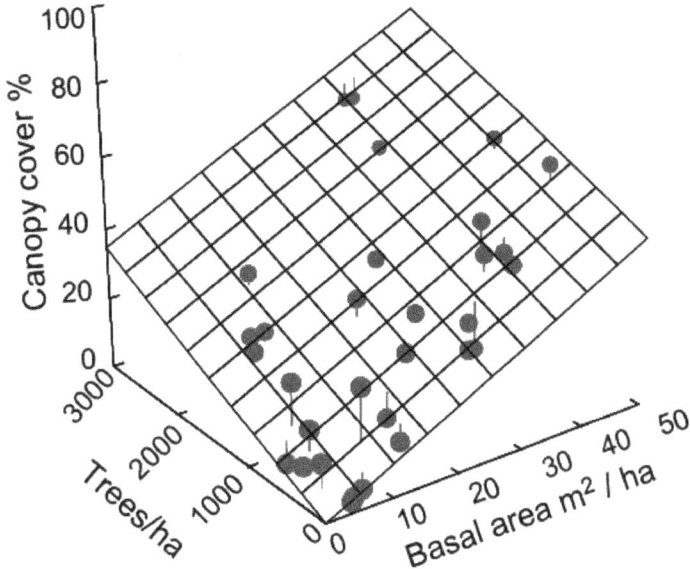

Figure 9. Lodgepole pine canopy cover as a linear function of basal area and tree density estimated with LAD regression for 31 sample stands.

A polynomial regression on a single independent variable, its square, its cube, and so on can be performed by including in the data file a column containing the square, cube, and so on of the independent variable as well as the original independent and dependent variable. We expect the user to have access to a commercial statistical package to perform these data transformations and graph results outside of Blossom. USE the file FRASERF.DAT and enter the following LAD command:

>LAD SCC = APIEN + PIENPHA + APIEN2

to estimate the model used in Cade (1997), where canopy cover of Engelmann spruce is predicted as a function of basal area (APIEN), basal area^2 (APIEN2), and stem density (PIENPHA). The results are below and the regression surface is plotted in Fig. 10:

```
            Least Absolute Deviation Regression (LAD)

Data Used
   Data file: FRASERF.DAT

LAD Regression:
     SCC=APIEN+PIENPHA+APIEN2
Results
   Number of observations: 31
       Dependent Variable: SCC

   Independent variables          Regression coefficients
```

```
APIEN                           1.58247874202118
PIENPHA                         0.842260983423992E-002
APIEN2                         -0.301945767106048E-001

Number of iterations: 5
Sum of absolute values of the residuals: 85.8394444292458
                Solution: SUCCESSFUL
```

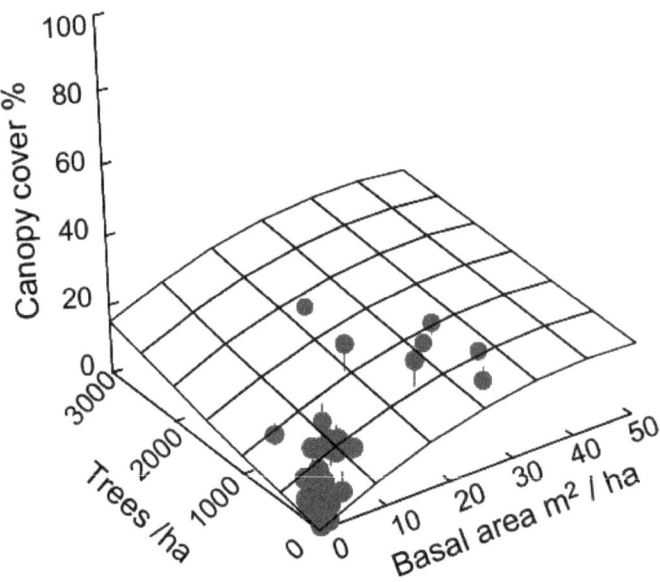

Figure 10. Engelmann spruce canopy cover as a quadratic function of basal area and linear function of tree density estimated with LAD regression for 31 sample stands.

Here the quadratic curvature implied by use of basal area^2 can be tested with the HYPOTHESIS command to test whether the addition of the squared term yielded an improvement in fit. This is equivalent to testing the full model specified above against a reduced model that doesn't include the term (APIEN2) for basal area^2. This is done by algebraically specifying the reduced parameter null model in the HYPOTHESIS command after the LAD command for the full parameter alternative model has been specified:

>HYPOTHESIS SCC = APIEN + PIENPHA / DP NPERM = 10000

Here are the results for the HYPOTHESIS command where we optionally have selected the double permutation scheme because our null hypothesized model is constrained through the origin:

```
                  Least Absolute Deviation Regression (LAD)
        Hypothesis test, drop p - q - 1 zero residuals, with double permutation

Data Used
   Data file: FRASERF.DAT

HYPOTHESIS Regression:
    SCC=APIEN+PIENPHA
Results
   Number of observations: 31
       Dependent Variable: SCC

   Independent variables          Regression coefficients
   APIEN                          0.613499653715770
   PIENPHA                        0.135282877749337E-001

   Number of iterations: 3
   Sum of absolute values of the residuals: 99.3797889162779
              Solution: SUCCESSFUL

Regression Evaluation:
   LAD Model:
      SCC=APIEN+PIENPHA+APIEN2
   Versus Hypothesis Model:
      SCC=APIEN+PIENPHA

Test Summary
   Number of permutations: 10000
       Random Number Seed: 3211532
   Observed Test Statistic: 0.157740355579689
   P-value of variables in full model but not in reduced model:
                      0.0140000000000000
```

The results indicate that the coefficient for the quadratic basal area term differs from zero with $P = 0.014$. Here both double permutation and dropping of all but 1 of the zero residuals under the null model were implemented because the null model includes 2 parameters but no intercept. If we had not used the double permutation option (/DP) and not deleted one of the zero residuals associated with the 2 parameters fit under the null model, then the P-value would be slightly smaller (0.0091) as in Cade (1997). The double permutation and dropping of zero residuals usually will increase the size of P-values slightly.

A goodness-of-fit measure for regression models is often a useful summary statistic. It is possible to compute a LAD coefficient of determination for the full model with reference to some reduced model (usually that specifies just an intercept term) by estimating the full model and obtaining the sums of absolute deviation (call it *SAF*), then estimating the reduced parameter model and obtaining its sum of absolute deviations (call it *SAR*), and computing the coefficient of determination $R^1 = 1 - (SAF/SAR)$ (Cade and Richards 1996, Cade 1997). This can be extended to any selected regression quantile by replacing the sums of absolute deviations in the formula above with the sum of weighted absolute deviations minimized by regression quantiles (Koenker and Machado 1999). We've already obtained the sums for the full parameter model, SCC = APIEN + PIENPHA + APIEN2 as *SAF* = 85.839, so to obtain them for the reduced parameter model:

LAD SCC = CONSTANT


```
                    Least Absolute Deviation Regression (LAD)

 Data Used
    Data file: FRASERF.DAT

 LAD Regression:
     SCC=CONSTANT
 Results
    Number of observations: 31
         Dependent Variable: SCC

    Independent variables           Regression coefficients
    CONSTANT                        10.0000000000000

    Number of iterations: 1
    Sum of absolute values of the residuals: 230.000000000000
               Solution: SUCCESSFUL
```

yields a sum, SAR = 230.0 for and, thus, the coefficient of determination R^1 = 1 - (85.839 / 230.000) = 0.627. This is interpreted as the model with variables APIEN, PIENPHA, and APIEN2 yield estimates of conditional medians of LCC with a 63% reduction in sum of absolute deviations compared to the model that is just a simple estimate of the median of LCC.

It is possible to specify greater or fewer permutations for calculating probabilities by specifying number of permutations as an option after either the test option for LAD command or as an option after HYPOTHESIS command. For example:

```
>USE FRASERF.DAT
>LAD LCC = APICO + PICOPHA / TEST NPERM = 10000 SEED = 123456
```

will test all slope parameters equal to zero using 10,000 permutations of y. Manly (1991) summarizes recommendations on number of permutations to use in Monte Carlo sampling procedures. More is better but comes at increased computational cost. Specifying the random number seed is done with the SEED = *num* option.

It is important to recognize that the LAD regression model (and generalization to regression quantiles discussed below) can be extended to any linear model design that might be estimated with OLS regression, including various variable transformations, and mixtures of continuous independent variables with indicator variables for categorical predictors. Extensive examples are in Mielke and Berry (2001). Indeed it is possible to use LAD regression for linear model analyses of multifactorial experimental designs, where the focus is on estimating changes in conditional medians rather than estimating changes in conditional means as typically done with OLS regression (Cade and Richards 1996, Mielke and Berry 2001).

As an example, consider the soap production example from Cade and Richards (1996), where soap scrap (v) is modeled as a linear function of production line speed (X_1) and an indicator variable $X_2 = 1$ for production line 1 and $X_2 = 0$ for production line 2 (Fig 11). We are interested in testing whether the rates of change in soap scrap (v) as a function of line speed (X_1) differs by production line (X_2), which requires that we estimate a model with an interaction term ($X_1 X_2$).

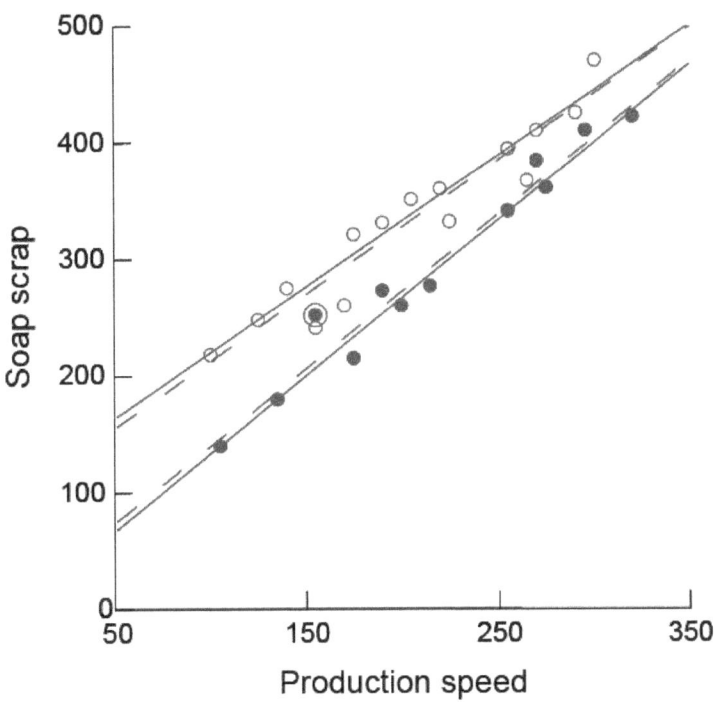

Figure 11. Soap scrap as a linear function of production speed for line1 (open circles) and line 2 (solid circles). Circled solid circle is an outlying value. Solid lines are LAD estimates and dashed lines are OLS estimates.

Open the data file NETER365.DAT and estimate the full parameter model with the interaction term specified

    LAD SOAP = CONSTANT + SPEED + LINE + LXS

where the LXS is a column variable created by multiplying SPEED times LINE across all observations. Here are the results:

```
              Least Absolute Deviation Regression (LAD)

Data Used
    Data file: NETER365.DAT

LAD Regression:
      SOAP = CONSTANT + SPEED + LINE + LXS
Results
```

```
Number of observations: 27
    Dependent Variable: SOAP

Independent variables              Regression coefficients
CONSTANT                           -0.319744231092045E-013
SPEED                              1.33333333333333
LINE                               107.615384615385
LXS                                -0.210256410256411
Number of iterations: 5
Sum of absolute values of the residuals: 389.435897435897
             Solution: SUCCESSFUL
```

Interpretation of the parameter estimates is identical to the interpretation for linear models estimated by OLS regression: the CONSTANT term is the intercept and the SPEED term (X_1), is the slope for the regression of soap scrap on line speed for line 2, the LINE term (X_2) is the difference between intercepts for the regressions for line 1 and line 2, and the LXS interaction term (X_1X_2) is the difference between slopes for the regressions for lines 1 and 2. We want to test the null hypothesis that the estimated interaction term is equal to zero, i.e., differences in slopes equals zero, by specifying the reduced parameter null model in the HYPOTHESIS command:

HYPOTHESIS SOAP = CONSTANT + SPEED + LINE/ NPERM = 10000

The results below indicated that there was moderate evidence ($P = 0.046$) that the estimated difference in slopes of -0.21 for the interaction term LXS was not equal to zero. Note that without dropping 2 of the 3 zero residuals in the null hypothesized model the P-value would be slightly smaller at $P = 0.031$.

```
              Least Absolute Deviation Regression (LAD)
              Hypothesis test, drop p - q - 1 zero residuals

Data Used
   Data file: NETER365.DAT

HYPOTHESIS Regression:
    SOAP =CONSTANT + SPEED + LINE
Results
   Number of observations: 27
      Dependent Variable: SOAP

   Independent variables              Regression coefficients
   CONSTANT                           39.2500000000000
   SPEED                              1.18333333333333
   LINE                               60.4166666666667

   Number of iterations: 6
   Sum of absolute values of the residuals: 451.750000000000
             Solution: NON-UNIQUE

Regression Evaluation:
   LAD Model:
      SOAP = CONSTANT + SPEED + LINE + LXS
   Versus Hypothesis Model:
      SOAP =CONSTANT + SPEED + LINE

Test Summary
```

```
Number of permutations: 10000
     Random Number Seed: 3231063
  Observed Test Statistic: 0.160011193047143
P-value of variables in full model but not in reduced model:
                    0.045500000000000
```

Confidence intervals on parameters in a LAD regression model can be constructed by inverting the hypothesis testing process in an iterative fashion. This is accomplished by recognizing that testing for nonzero values of parameters in null hypotheses only requires a linear transformation of the dependent variable, y. For example, for the H_0: $\beta_1 = \lambda$, where λ is some hypothesized value of the parameter, you transform y to, say z, by $z = y - \lambda X_1$. The transformed values of the dependent variable, z, are then substituted for y in the regression model and estimation and hypothesis testing of the null H_0: $\beta_1 = 0$ proceed as before. Cade and Richards (1996) describe in more general matrix notation how you accomplish this linear transformation for multiple parameters. Note that the formula defaults to what is done automatically when we test null hypotheses that parameters, λ, equal zero. The complication that arises in implementing this procedure for a $(1 - \alpha)$% confidence interval is that you must iterate through many possible values of λ to define the bounds on the set of values of λ with $P \geq \alpha$ for H_0: $\beta_1 = \lambda$. This can require many transformations of y, estimation with LAD, and testing the null hypothesis with the HYPOTHESIS command.

As an example of constructing confidence intervals, return to the model of lodgepole pine canopy cover as a function of pine basal area and stem density (Cade 1997). Endpoints of the 95% confidence interval for the basal area parameter ($b_1 = 0.935$) were given as 0.81 - 1.05 in Cade (1997). This means that the transformations LCC - 0.81(APICO), call it $Z81$, and LCC - 1.05(APICO), call it $Z105$, should have approximate $P = 0.05$ when $Z81$ and $Z105$ are substituted for LCC in the regression model that includes APICO (basal area) and PICOPHA (tree density) as predictors for the partial model hypothesis of APICO. Any transformation of LCC by values between 0.81 and 1.05 ought to yield $P > 0.05$ and any outside of this interval ought to yield $P \leq 0.05$. Minor discrepancies can occur, of course, because of the resampling variation inherent in Monte Carlo procedures and because of discreteness in the permutation distribution. We try and make the resampling error as small as possible by using a large number of permutations (NPERM $\geq$ 10,000). The file FRASERF.DAT includes the transformations $Z81$ and $Z105$, as well as $Z90$ = LCC - 0.90(APICO) and $Z50$ = LCC - 0.50(APICO). Here we know that the interval presented in Cade (1997) is slightly narrower than expected when the more recently developed double permutation scheme (Cade 2005, Cade and Richards In press) is used because the null model is constrained through the origin. To run the hypothesis test corresponding to the null model that the parameter for APICO equals 0.81, issue the commands:

```
>USE FRASERF.DAT
>LAD Z81 = APICO + PICOPHA
>HYPOTHESIS Z81 = PICOPHA/ NPERM = 10000 DP
```

The output indicates $P = 0.0562$, well within the Monte Carlo resampling variation of 0.05 as it should be and only slightly larger than $P = 0.0523$ obtained without the double permutation scheme.

```
                  Least Absolute Deviation Regression (LAD)

Data Used
   Data file: FRASERF.DAT

LAD Regression:
    Z81=APICO+PICOPHA
Results
   Number of observations: 31
      Dependent Variable: Z81

   Independent variables           Regression coefficients
   APICO                           0.124738765015896
   PICOPHA                         0.115723762503457E-001

   Number of iterations: 4
   Sum of absolute values of the residuals: 127.511282247993
            Solution: SUCCESSFUL

=====================================================================
                  Least Absolute Deviation Regression (LAD)
                  Hypothesis test with double permutation

Data Used
   Data file: FRASERF.DAT

HYPOTHESIS Regression:
    Z81=PICOPHA
Results
   Number of observations: 31
      Dependent Variable: Z81

   Independent variables           Regression coefficients
   PICOPHA                         0.128218706607143E-001

   Number of iterations: 1
   Sum of absolute values of the residuals: 139.983715196277
            Solution: SUCCESSFUL

Regression Evaluation:
   LAD Model:
     Z81=APICO+PICOPHA
   Versus Hypothesis Model:
     Z81=PICOPHA

Test Summary
    Number of permutations: 10000
        Random Number Seed: 3245102
   Observed Test Statistic: 0.978143480984412E-001
   P-value of variables in full model but not in reduced model:
                    0.05620000000000
```

Similarly, we can run the hypothesis test corresponding to the null that the parameter for APICO equals 0.50 by issuing the commands:

```
>USE FRASERF.DAT
>LAD Z50 = APICO + PICOPHA
>HYPOTHESIS Z50 = PICOPHA/ NPERM = 10000 DP
```

The output here indicates the null hypothesis that the parameter equals 0.50 has $P = 0.0001$, much smaller than 0.05 so that this hypothesized parameter value must be outside the 95% confidence interval.

```
                Least Absolute Deviation Regression (LAD)

Data Used
    Data file: FRASERF.DAT

LAD Regression:
    Z50=APICO+PICOPHA
Results
    Number of observations: 31
        Dependent Variable: Z50

    Independent variables          Regression coefficients
    APICO                          0.434738765013986
    PICOPHA                        0.115723762499917E-001

    Number of iterations: 4
    Sum of absolute values of the residuals: 127.511282246895
                Solution: SUCCESSFUL

==========================================================================


                Least Absolute Deviation Regression (LAD)
                   Hypothesis test with double permutation

Data Used
    Data file: FRASERF.DAT

HYPOTHESIS Regression:
    Z50=PICOPHA
Results
    Number of observations: 31
        Dependent Variable: Z50

    Independent variables          Regression coefficients
    PICOPHA                        0.156154562384342E-001

    Number of iterations: 1
    Sum of absolute values of the residuals: 211.176442288515
                Solution: SUCCESSFUL

Regression Evaluation:
    LAD Model:
        Z50=APICO+PICOPHA
```

```
Versus Hypothesis Model:
   Z50=PICOPHA

Test Summary
     Number of permutations: 10000
         Random Number Seed: 3254170
   Observed Test Statistic: 0.656139273069367
   P-value of variables in full model but not in reduced model:
                          0.100000000000000E-003
```

Presently, hypothesized values of the parameter and their transformations must be made iteratively by successive approximation, i.e., guess at values, compute the *P*-values, and then based on the size of the *P*-value successively move towards larger or lower values until you have values with $P = \alpha$, which define the confidence interval endpoints. This can require 20 or more iterations depending on how close your initial choice of hypothesized parameter values are to the final values. It is possible to use asymptotic procedures described in Birkes and Dodge (1993) to help pick initial values for confidence interval endpoints that might be close to those obtained by the iterative permutation testing process.

Regression Quantiles

The QUANT = *num* | ALL option of the LAD regression command fits any specified conditional quantile as a linear regression model. LAD regression is the 0.50 (50th percentile) regression quantile. Various regression quantiles, e.g, 0.05, 0.10, 0.25, 0.50, 0.75, 0.90, 0.95 (i.e., 5th, 10th, 25th, 75th, 90th and 95th percentiles), can be estimated to examine linear trends in a dependent variable (*y*) as a function of one or more independent variables (*X*). Selecting QUANT = ALL will yield all possible quantile regression estimates. If there is little variation in the errors across the independent variables (homogeneous errors), the regression quantiles will have similar slopes but different intercepts. However, if the errors are heterogeneous across the independent variables, then slopes and intercepts can differ greatly (Cade and Richards 1996, Terrell et al. 1996, Cade et al. 1999 Koenker and Machado 1999). Regression quantiles, thus, provide a way of modeling rates of change associated with heterogeneous variation in linear models without having to specify a functional link between conditional measures of means and and variances. Regression quantiles are especially useful when the consequences of over and under prediction differ in a linear model. Cade and Noon (2003) present a primer on quantile regression for ecologists.

In studies of ecological limiting factors it is often expected that important measured processes operate as constraints on the response distribution (*y*) and, thus, we may focus on estimating regression quantiles associated with the upper percentiles (e.g., 90 - 99th) of the dependent variable, i.e., rates of change estimated are along the upper boundary of the distribution as it changes across the independent variables (Terrell et al. 1996, Cade et al. 1999, Haire et al. 2000, Cade and Guo 2000). Rates of change in the responses below the boundary constraint may be lower because of the impact of unmeasured processes (Cade et al. 1999). Many ecological processes can be considered constraints on responses, where rates of change estimated with regression quantiles for upper percentiles might yield new insights. Examples include animal responses to habitat, self-thinning in plants, algal productivity as a function of limiting nutrients, animal abundance and body size relations in macroecology, comparisons of local and regional species diversity, plant productivity as a function of species diversity, and competition field experiments. Estimating rates of change for endpoints of some interval of quantiles (e.g., 10th and 90th percentiles) also provides a flexible way to estimate prediction intervals for responses without resorting to untenable distributional assumptions.

Returning to the soap production example, after USEing the file NETER365.DAT, issue the following command:

>LAD SOAP = CONSTANT + SPEED + LINE + LXS/QUANT = 0.50

The output indicates that the coefficients estimated are identical to those above without the QUANT = 0.5 option, because the 0.5 quantile is LAD regression. Notice also that both the sum of absolute deviations minimized in LAD regression and the sum of weighted absolute deviations minimized in regression quantiles are reported. The weights used when minimizing sums of absolute deviations in regression quantiles are τ for positive residuals and $1 - \tau$ for zero and negative residuals, where $0 \le \tau \le 1$ is the selected quantile with QUANT = num. Thus, in this example the sum of weighted absolute deviations is exactly half the sum of absolute deviations.

```
                    Quantile Regression

Data Used
    Data file: NETER365.DAT

0.50 Quantile Regression:
     SOAP = CONSTANT + SPEED + LINE +LXS
Results
    Number of observations: 27
       Dependent Variable: SOAP

               For Quantile = 0.50
    Independent variables          Regression coefficients
    CONSTANT                       -0.319744231092045E-013
    SPEED                          1.33333333333333
    LINE                           107.615384615385
    LXS                            -0.210256410256411

    Number of iterations: 5
    Sum of absolute values of the residuals: 389.435897435897
    Weighted sum of the absolute deviations: 194.717948717949
              Solution: SUCCESSFUL
```

It is possible to test a full versus a reduced parameter regression quantile model with the default TEST and HYPOTHESIS options as in the LAD regression command, where the test statistic is identical in computation as for LAD except that the simple sum of absolute deviations are replaced with the sum of weighted absolute deviations (Cade 2005, Cade and Richards In Press). Validity of hypothesis tests for regression quantiles using this test statistic requires the same assumption of independent, identical error distributions as for LAD regression. However, we expect most applications of regression quantiles to be made when it is unreasonable to assume homogeneous variation across the independent variables, i.e., the identical error distribution assumption is violated. Therefore, we have included the regression quantile rank score test (Koenker 1994, Koenker and Machado 1999), its asymptotic *P*-value approximation with a Chi-square distribution , and a permutation approximation that makes use of the permutation test for OLS regression. Type I errors of the regression quantile rank score test are less sensitive to heterogeneous error distributions because the statistic is based on the sign of the residuals from the reduced parameter null model and not their size. However, as Cade (2003) and Cade et al. (2005b) make abundantly clear, valid Type I error rates often will require appropriate weighted estimates and test statistics. This quantile rank score test is implemented with the option / RANKSCORE given with the HYPOTHESIS command.

As an example, consider the acorn production data as related to oak (*Quercus spp.*) forest characteristics (Schroeder and Vangilder 1997) as analyzed with regression quantiles by Cade et al. (1999). We will estimate 0.10 and 0.90 (10th and 90th percentiles) regression quantiles of annual acorn biomass (kg/ha) as a function of a forest suitability index based on canopy cover and number of oak species (Schroeder and Vangilder 1997). USE the data file ACORN.DAT and issue the command for a 0.10 regression quantile:

>LAD WTPERHA = CONSTANT + OAKCCSI/ QUANT = 0.10

The command then is issued to test the hypothesis that the slope for the 0.10 quantile equals zero with the rank score test:

>HYPOTHESIS WTPERHA = CONSTANT/ RANKSCORE NPERM = 10000

The output indicates that the estimated slope for the 0.10 regression quantile (21.8) likely differs from zero (*P* = 0.012).

```
                    Quantile Regression

Data Used
   Data file: ACORN.DAT

0.10 Quantile Regression:
     WTPERHA=CONSTANT+OAKCCSI
Results
   Number of observations: 43
       Dependent Variable: WTPERHA

             For Quantile = 0.10
   Independent variables          Regression coefficients
   CONSTANT                       2.44020411434225
   OAKCCSI                        21.7718847844891

   Number of iterations: 2
   Sum of absolute values of the residuals: 1526.32980484628
   Weighted sum of the absolute deviations: 173.730348688158
             Solution: SUCCESSFUL
```

```
========================================================================

                      Quantile Regression
                 Hypothesis test of Rank Score

Data Used
   Data file: ACORN.DAT

0.10 Quantile HYPOTHESIS Regression:
     WTPERHA=CONSTANT
Results
   Number of observations: 43
       Dependent Variable: WTPERHA

             For Quantile = 0.10
   Independent variables        Regression coefficients
   CONSTANT                     12.8247400000000

   Number of iterations: 1
   Sum of absolute values of the residuals: 1737.33740600000
   Weighted sum of the absolute deviations: 194.228089400000
               Solution: SUCCESSFUL

Regression Evaluation:
   0.10 Quantile Regression Model:
     WTPERHA=CONSTANT+OAKCCSI
   Versus Hypothesis Model at Quantile 0.10:
     WTPERHA=CONSTANT

Test Summary
   Number of permutations: 10000
        Random Number Seed: 3274240
   Observed Rank Score Test Statistic: 0.184214175628134
         P-value of Rank Score Test: 0.960000000000000E-002
     Asymptotic Rank Score Statistic: 6.32603175145283
         (Distributed as Chi-square with degrees of
          freedom equal to difference in number of
          parameters between full and reduced models.)
       P-Value of Asymptotic RS Stat: 0.118978245412483E-001
```

Similarly, we can estimate the 0.90 regression quantile for the same functional relation by issuing the command:

>LAD WTPERHA = CONSTANT + OAKCCSI/ QUANT = 0.90

followed by the command:

>HYPOTHESIS WTPERHA = CONSTANT/ RANKSCORE NPERM = 10000

The output for the 0.90 regression quantile indicates that the rate of change of acorn biomass with the suitability index is 5 times greater (102.3) at the 90th percentile of the distribution compared to the 10th percentile of the distribution (Fig 12). Clearly, there is heterogeneous variation in the acorn biomass changes across the acorn suitability index, with only larger biomass occurring at higher values of the suitability index. The estimated slope of the 0.90

regression quantile also likely differs from zero ($P = 0.040$). Here because of the heterogeneity, improved Type I error rates could be obtained by using weighted estimates with the rank score tests.

```
                        Quantile Regression

Data Used
   Data file: ACORN.DAT

0.90 Quantile Regression:
    WTPERHA=CONSTANT+OAKCCSI
Results
   Number of observations: 43
      Dependent Variable: WTPERHA

            For Quantile = 0.90
   Independent variables          Regression coefficients
   CONSTANT                       14.4485704551128
   OAKCCSI                        102.338029544887

   Number of iterations: 3
   Sum of absolute values of the residuals: 1722.33277425585
   Weighted sum of the absolute deviations: 268.539600522219
            Solution: SUCCESSFUL

======================================================================


                        Quantile Regression
                    Hypothesis test of Rank Score

Data Used
   Data file: ACORN.DAT

0.90 Quantile HYPOTHESIS Regression:
    WTPERHA=CONSTANT
Results
   Number of observations: 43
      Dependent Variable: WTPERHA

            For Quantile = 0.90
   Independent variables          Regression coefficients
   CONSTANT                       89.9235100000000

   Number of iterations: 1
   Sum of absolute values of the residuals: 1951.43425600000
   Weighted sum of the absolute deviations: 324.058897600000
            Solution: SUCCESSFUL

Regression Evaluation:
   0.90 Quantile Regression Model:
     WTPERHA=CONSTANT+OAKCCSI
   Versus Hypothesis Model at Quantile 0.90:
     WTPERHA=CONSTANT

Test Summary
    Number of permutations: 10000
        Random Number Seed: 3288861
```

```
Observed Rank Score Test Statistic: 0.115100869658327
        P-value of Rank Score Test: 0.352000000000000E-001
    Asymptotic Rank Score Statistic: 4.19761909151131
        (Distributed as Chi-square with degrees of
        freedom equal to difference in number of
        parameters between full and reduced models.)
      P-Value of Asymptotic RS Stat: 0.404807769147461E-001
```

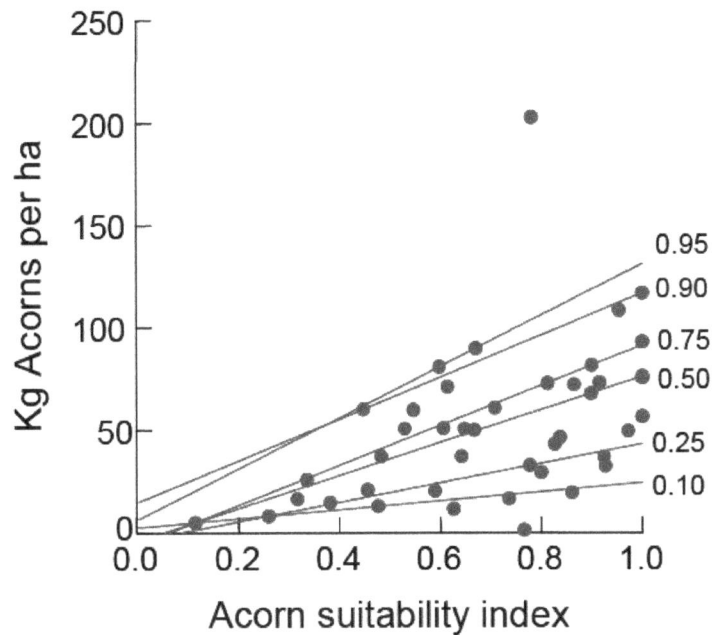

Figure 12. Average annual biomass of acorns and acorn suitability indices based on oak forest characteristics in 43 0.2-ha sample plots in Missouri. Solid lines are estimates for 6 selected regression quantiles.

Estimates for other regression quantiles can be obtained by changing the value used in the option QUANT = *num*. Note that the *P*-values approximated by the permutation evaluation of the rank score tests are similar to those produced by the asymptotic Chi-square distributional approximation (uses a Chi-square distribution with degrees of freedom equal to difference in number of parameters in full versus reduced models). Although the permutation *P*-values are slightly smaller than those for the asymptotic Chi-square approximation, the differences may be attributable just to the resampling error associated with the Monte Carlo approximation. Simulation research in Cade (2003) and Cade et al. (2005b) established that the permutation version of the rank score test maintains valid Type I error rates at more extreme quantiles (τ) with smaller *n* than does the Chi-square distributional approximation.

Confidence intervals based on the regression quantile rank score statistic can be formed by a process identical to that described above for LAD regression. However, if you want to use the asymptotic Chi-square approximation of *P*-values for computing confidence intervals, there are fast implementations in linear programming algorithms available for S-Plus, R, and SAS (Koenker 1994, Cade et al. 1999, Koenker and Machado 1999).

A multiple regression quantile example is provided by Cade et al. (1999), where glacier lily (*Erythronium grandiflorum*) seedlings are linearly related to the number of flowers and an index of rockiness in *n* = 256 contiguous 2 × 2 m quadrats (Fig. 13).

To estimate the 95th regression quantile model issue the following commands:

```
>USE LILY.DAT
>LAD SEEDLINGS = CONSTANT + FLOWERS + ROCKINESS/ QUANT = 0.95
```

and obtain the following output:

```
                    Quantile Regression
Data Used
   Data file: lily.dat

0.95 Quantile Regression:
    SEEDLINGS = CONSTANT + FLOWERS +ROCKINESS
Results
   Number of observations: 256
        Dependent Variable: SEEDLINGS

            For Quantile = 0.95
   Independent variables            Regression coefficients
   CONSTANT                         20.2991556091677
   FLOWERS                          0.850422195416164E-001
   ROCKINESS                        -0.898673100120627E-001

   Number of iterations: 5
   Sum of absolute values of the residuals: 3800.25995174910
   Weighted sum of the absolute deviations: 272.780971049457
                Solution: SUCCESSFUL
```

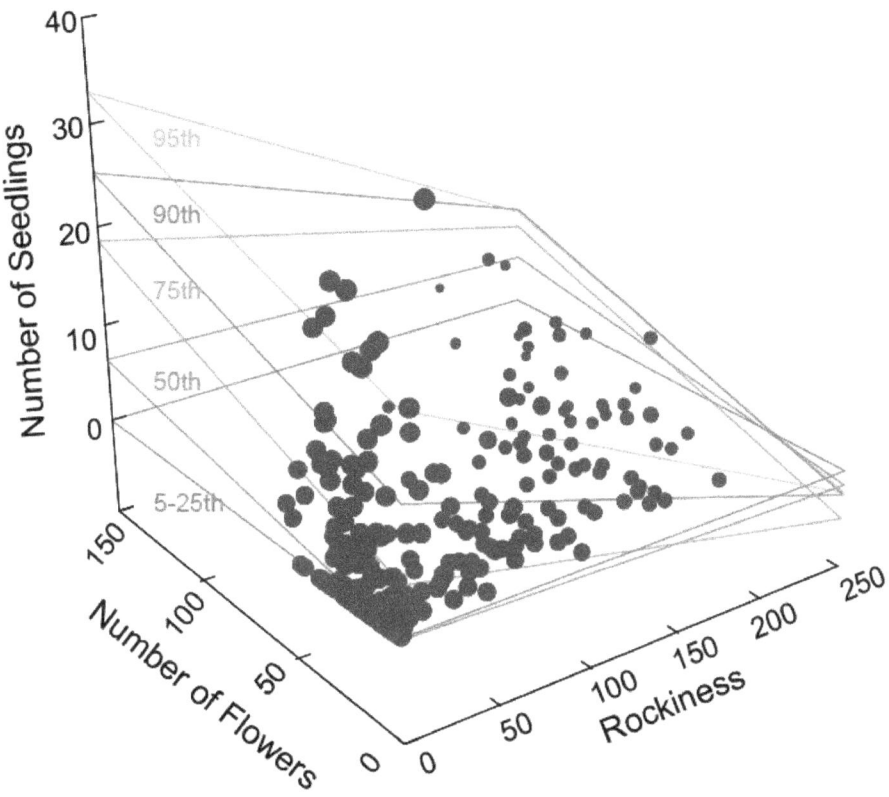

Figure 13. Glacier lily seedling counts, lily flower numbers, and rockiness index for 256 2 × 2 m quadrats in subalpine meadow of western Colorado. Surfaces are for selected regression quantile estimates (1 outlying count of 72 seedlings is not plotted).

The estimates indicate a 0.085 increase in seedling numbers with each increase in flower numbers at a given level of rockiness, and a decrease of 0.090 of seedling numbers with each increase in unit of the rockiness index. We can test that these parameters jointly are equal to zero by comparing the full parameter model above with the reduced parameter model having just an intercept by the command:

>HYPOTHESIS SEEDLINGS = CONSTANT/ RANKSCORE NPERM = 10000

The output indicates some evidence that at least one of the parameters is unlikely to equal zero ($P = 0.030$ for asymptotic approximation and $P = 0.028$ for permutation approximation).

```
                    Quantile Regression
                Hypothesis test of Rank Score

Data Used
    Data file: lily.dat
```

```
0.95 Quantile HYPOTHESIS Regression:
     SEEDLINGS = CONSTANT
Results
   Number of observations: 256
       Dependent Variable: SEEDLINGS


             For Quantile = 0.95
   Independent variables          Regression coefficients
   CONSTANT                       16.0000000000000

   Number of iterations: 1
   Sum of absolute values of the residuals: 3377.00000000000
   Weighted sum of the absolute deviations: 301.150000000001
             Solution: SUCCESSFUL

Regression Evaluation:
   0.95 Quantile Regression Model:
      SEEDLINGS = CONSTANT + FLOWERS +ROCKINESS
   Versus Hypothesis Model at Quantile 0.95:
      SEEDLINGS = CONSTANT

Test Summary
    Number of permutations: 10000
        Random Number Seed: 3314138
   Observed Rank Score Test Statistic: 0.285678029923178E-001
           P-value of Rank Score Test: 0.278000000000000E-001
       Asymptotic Rank Score Statistic: 7.01667809940338
         (Distributed as Chi-square with degrees of
          freedom equal to difference in number of
          parameters between full and reduced models.)
         P-Value of Asymptotic RS Stat: 0.299466129878649E-001
```

We can test each of the parameters individually by issuing the series of commands:

>HYPOTHESIS SEEDLINGS = CONSTANT + FLOWERS / RANKSCORE
      NPERM = 10000
>HYPOTHESIS SEEDLINGS = CONSTANT + ROCKINESS/ RANKSCORE
      NPERM = 10000

The output indicates stronger evidence that the parameter for ROCKINESS does not equal zero ($P = 0.041$) than for the parameter for FLOWERS ($P = 0.079$).

```
                  Quantile Regression
               Hypothesis test of Rank Score

Data Used
   Data file: lily.dat

0.95 Quantile HYPOTHESIS Regression:
     SEEDLINGS = CONSTANT+FLOWERS
Results
   Number of observations: 256
       Dependent Variable: SEEDLINGS
```

```
            For Quantile = 0.95
    Independent variables           Regression coefficients
    CONSTANT                        18.5818181818182
    FLOWERS                         -0.727272727272729E-001


    Number of iterations: 4
    Sum of absolute values of the residuals: 3460.49090909091
    Weighted sum of the absolute deviations: 296.062727272727
            Solution: SUCCESSFUL

Regression Evaluation:
    0.95 Quantile Regression Model:
      SEEDLINGS = CONSTANT + FLOWERS +ROCKINESS
    Versus Hypothesis Model at Quantile 0.95:
      SEEDLINGS = CONSTANT+FLOWERS


Test Summary
      Number of permutations: 10000
          Random Number Seed: 3325381
    Observed Rank Score Test Statistic: 0.171472162215515E-001
            P-value of Rank Score Test: 0.365000000000000E-001
      Asymptotic Rank Score Statistic: 4.16986218022584
          (Distributed as Chi-square with degrees of
           freedom equal to difference in number of
           parameters between full and reduced models.)
        P-Value of Asymptotic RS Stat: 0.411491467475027E-001


========================================================================

                    Quantile Regression
              Hypothesis test of Rank Score

Data Used
    Data file: lily.dat

0.95 Quantile HYPOTHESIS Regression:
      SEEDLINGS = CONSTANT+ROCKINESS
Results
    Number of observations: 256
        Dependent Variable: SEEDLINGS


            For Quantile = 0.95
    Independent variables           Regression coefficients
    CONSTANT                        22.0000000000000
    ROCKINESS                       -0.652173913043478E-001


    Number of iterations: 4
    Sum of absolute values of the residuals: 3918.43478260870
    Weighted sum of the absolute deviations: 278.565217391304
            Solution: SUCCESSFUL

Regression Evaluation:
    0.95 Quantile Regression Model:
      SEEDLINGS = CONSTANT + FLOWERS +ROCKINESS
    Versus Hypothesis Model at Quantile 0.95:
      SEEDLINGS = CONSTANT+ROCKINESS


Test Summary
      Number of permutations: 10000
          Random Number Seed: 3330654
    Observed Rank Score Test Statistic: 0.125367654441921E-001
            P-value of Rank Score Test: 0.765000000000000E-001
```

```
Asymptotic Rank Score Statistic: 3.09446818237270
      (Distributed as Chi-square with degrees of
       freedom equal to difference in number of
       parameters between full and reduced models.)
      P-Value of Asymptotic RS Stat: 0.785588175421340E-001
```

Both *P*-values are consistent with the 90% confidence intervals given in Cade et al. (1999) that did not overlap zero for either variable. Note that the permutation *P*-values are slightly smaller than the Chi-square distribution approximation. The confidence intervals in Cade et al. (1999) were based on inverting the asymptotic Chi-square distribution approximation of the rank score statistic as part of the linear programming solution for regression quantiles that are available for S-Plus (see Ecological Archives E080-001 for these routines). Because of the heterogeneity evident in this model, confidence intervals and rank score testing would be better based on weighted estimates (Cade et al. 2005b)

The use of all quantile regression estimates and weighting is provided for an example relating Lahontan cutthroat trout (*Oncorhynchus clarki henshawi*) numbers per meter of stream to stream width:depth ratio for *n* = 71 observations of streams across years in Nevada (Dunham et al. 2002, Cade 2005, Cade et al. 2005b). The scatter plot in Figure 14 (A) indicate moderate heterogeneity and some nonlinearity in the relationship. Dunham et al. (2002) chose to use a nonlinear model $y = \exp(\beta_0 + \beta_1 X_1 + \varepsilon)$ estimated in the linear scale by taking natural logarithms of both sides of the equation. Cade (2005), Cade et al. (2005b), and Cade and Richards (In press) also used weighted estimates, where the coefficients of the weight function $w = (1.310 - 0.0017X_1)^{-1}$ were estimated from the average pairwise differences (by using expected value obtained from multiresponse sequence procedure) between all possible quantile regression estimates for β_0 and for β_1 obtained by using the QUANT = ALL option:

```
USE LAHONTAN.DAT
LAD LNLCTM = CONSTANT +WIDRAT/ QUANT = ALL SAVE=ALLTROUT1.TXT
```

```
                      All Quantile Regressions

Data Used
   Data file: lahontan.dat

All Quantile Regressions From Command:
    LNLCTM=CONSTANT + WIDRAT

     Dependent Variable: LNLCTM
   Independent Variables:
        CONSTANT
        WIDRAT
     Number of Observations: 71
  Number of Model Parameters: 2
        Number of Solutions: 77
        Solution Result Was: SUCCESSFUL

 Full solution results are written to file lahontan.OUT
Output was appended to file "lahontan.OUT"
All quantile solutions saved in file: ALLTROUT1.TXT
```

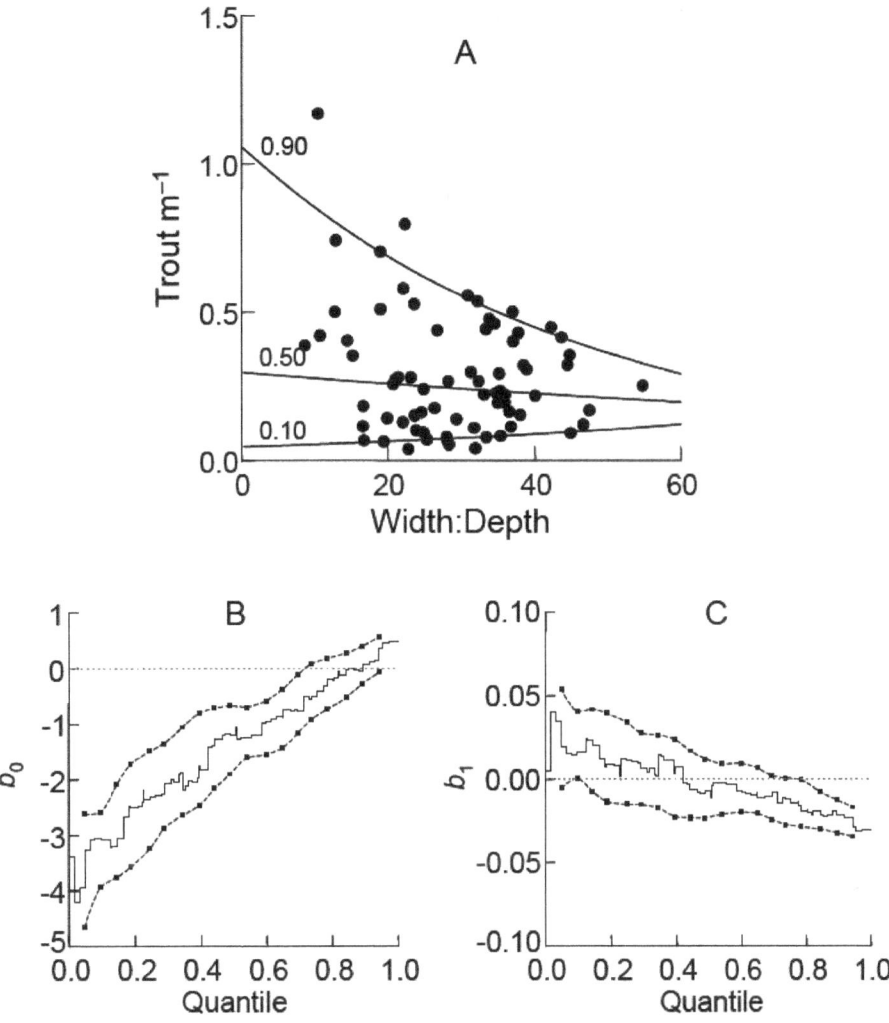

Figure 14. (A) Lahontan cutthroat trout m⁻¹ and width:depth ratios for small streams sampled 1993 to 1999 ($n = 71$); exponentiated estimates for 0.90, 0.50, and 0.10 regression quantiles for the weighted model $w(\ln y) = w(\beta_0 + \beta_1 X_1 + (\gamma_0 + \gamma_1 X_1)\varepsilon)$, $w = (1.310 - 1.017 X_1)^{-1}$. Solid lines in (B) and (C) are step functions for estimates of β_0 and β_1 by $\tau \in [0, 1]$ and dashed lines connect pointwise 90% confidence intervals for $\tau \in \{0.05, 0.10, 0.15, ..., 0.95\}$ based on inverting the double permutation test.

The file ALLTROUT1.TXT contains a row for each unique interval of quantiles, with column variables specifying the upper endpoint of the quantile interval (Quantile), the objective function minimized (ObjFuncSol is weighted sum of absolute deviations), the predicted value for that quantile at the mean of the independent variables (PredY_Xbar), and the parameter estimates (here, b_CONSTANT and b_WIDRAT). Plots of the parameter estimates by quantile suggested the linear location-scale (in log scale) form of heterogeneity was a reasonable approximation so that a single weight function could reasonably be applied to all quantiles. The empirical distribution plots for each parameter estimate by quantile in Figure 14 (B and C) were made from the weighted estimates by connecting the point estimates with an appropriate step function (Figure 14 B and C). The weighted estimates were made by multiplying all variables (LNLCTM, a column of 1's for the intercept, and WIDRAT) by the weights (WT) to form the variables WTLNLCTM, WT, and WTWIDRAT. The model was estimated as:

```
USE LAHONTAN.DAT
LAD WTLNLCTM = WT + WTWIDRAT/QUANT = ALL SAVE=ALLTROUT2.TXT
```

```
                    All Quantile Regressions

Data Used
   Data file: lahontan.dat

All Quantile Regressions From Command:
    WTLNLCTM=WT + WTWIDRAT

      Dependent Variable: WTLNLCTM
   Independent Variables:
      WT
      WTWIDRAT

      Number of Observations: 71
   Number of Model Parameters: 2
         Number of Solutions: 79
         Solution Result Was: SUCCESSFUL

 Full solution results are written to file lahontan.OUT
Output was appended to file "lahontan.OUT"
All quantile solutions saved in file: ALLTROUT2.TXT
```

Note that the variable WT (= 1 × WT) replaces the usual CONSTANT term because the weighted model requires that weights are multiplied by all independent variables including the column of 1's for the constant. The confidence intervals formed around the parameter estimates by quantiles in Figure 14 (B and C) were made by using the drop in dispersion permutation test with double permutation (because null models for weighted estimates were constrained through the origin). Cade and Richards (In press) formed 90% confidence intervals at quantiles = 0.05, 0.10, 0.15 ... 0.90, 0.95 by successive iteration of hypothesized values as explained for LAD regression starting on page 82. These intervals were only slightly narrower than intervals formed by inverting the permutation version or Chi-square distributional approximation of the rank score test (Cade et al. 2005b). Here, we provide an example of the hypothesis tests for the weighted 0.90 quantile regression estimates:

```
LAD WTLNLCTM = WT + WTWIDRAT/QUANT=0.90
HYP WTLNLCTM = WT/NPERM=100000 DP
HYP WTLNLCTM = WTWIDRAT/NPERM = 100000 DP
```

```
                    Quantile Regression

Data Used
   Data file: lahontan.dat

0.90 Quantile Regression:
     WTLNLCTM=WT + WTWIDRAT
Results
   Number of observations: 71
       Dependent Variable: WTLNLCTM

              For Quantile = 0.90
   Independent variables          Regression coefficients
   WT                             0.05762007758715407
   WTWIDRAT                       -0.02154147781141880

   Number of iterations: 2
   Sum of absolute values of the residuals: 82.41461016146272
   Weighted sum of the absolute deviations: 8.796595654366030
              Solution: SUCCESSFUL
Output was appended to file "lahontan.OUT"

>HYP WTLNLCTM=Wt/NPERM=100000 DP


                    Quantile Regression
            Hypothesis Test with Double Permutation

Data Used
   Data file: lahontan.dat

0.90 Quantile HYPOTHESIS Regression:
     WTLNLCTM=WT
Results
   Number of observations: 71
       Dependent Variable: WTLNLCTM

              For Quantile = 0.90
   Independent variables          Regression coefficients
   WT                             -0.6763883967527673

   Number of iterations: 1
   Sum of absolute values of the residuals: 80.29814001870930
   Weighted sum of the absolute deviations: 9.831009518510045
              Solution: SUCCESSFUL

Regression Evaluation:
   0.90 Quantile Regression Model:
     WTLNLCTM=WT + WTWIDRAT
   Versus Hypothesis Model at Quantile 0.90:
     WTLNLCTM=WT

Test Summary
     Number of permutations: 100000
          Random Number Seed: 46188336
     Observed Test Statistic: 0.1175925215603838
     P-value of variables in full model but not in reduced model:
                     0.002360000000000000
Output was appended to file "lahontan.OUT"


>HYP WTLNLCTM=WTWIDRAT/NPERM=100000 DP
```

```
                    Quantile Regression
            Hypothesis Test with Double Permutation

Data Used
   Data file: lahontan.dat

0.90 Quantile HYPOTHESIS Regression:
    WTLNLCTM=WTWIDRAT
Results
   Number of observations: 71
       Dependent Variable: WTLNLCTM

              For Quantile = 0.90
   Independent variables        Regression coefficients
   WTWIDRAT                     -0.02022132436065866

   Number of iterations: 1
   Sum of absolute values of the residuals: 81.25775342749679
   Weighted sum of the absolute deviations: 8.807900858607981
              Solution: SUCCESSFUL

Regression Evaluation:
   0.90 Quantile Regression Model:
     WTLNLCTM=WT + WTWIDRAT
   Versus Hypothesis Model at Quantile 0.90:
     WTLNLCTM=WTWIDRAT

Test Summary
    Number of permutations: 100000
        Random Number Seed: 46230889
     Observed Test Statistic: 0.001285179481489562
    P-value of variables in full model but not in reduced model:
                        0.7816900000000000
Output was appended to file "lahontan.OUT"
```

Note that both null hypothesized models above do not include a CONSTANT for a column of 1's because of the weighting scheme, so that the double permutation option DP was used to provide better Type I error rates. The output indicates a strong, nonzero slope but an intercept that doesn't differ from zero (in the log scale) for the 0.90 regression quantile. Notice that these results are consistent with the 90% CI which indicate nonzero slopes for quantiles ≥ 0.80 and nonzero intercepts for quantiles ≤ 0.70

The LAD Command Syntax

The LAD command can be used to fit a variety of least absolute deviation regressions. The HYPOTHESIS command allows the specification of reduced parameter LAD regression model to compare with the full parameter regression model specified in the main LAD command. The regressions are run and the tests performed upon entering the LAD and HYPOTHESIS commands. If the QUANT = *num* option is specified, all subsequent testing is done on the specified conditional quantile.

LAD *dep. var* = [CONSTANT +] *ind. var1* + *ind. var2* + ...

    [/TEST | NPERM = *num*| SEED = *num* | SAVE [= *file name*]
    | QUANT = *num* | *ALL*]

HYPOTHESIS *dep. var* = [CONSTANT +] *ind. var1* + *ind. var2* + ...

    [/NPERM = *num* | DP | SEED = *num* | RANKSCORE |
    SAVETEST [= *file name*]]

Items to be supplied by the user are given in lower case in italics. Items in square brackets are optional. The vertical line (|) can be read as "or" and separates different options that can be specified. They can be specified in any order. The single variable named on the left of the equal sign is the dependent variable. The independent variables are listed and separated by plus signs to indicate the form of the regression model. If the model is to include a constant (intercept term) the term CONSTANT must be placed right after the equal sign.

LAD options follow the slash (/) character. The TEST option causes the default test of all slope parameters equal to zero. The NPERM option allows the user to specify more or fewer permutations than the default of 5,000 used in approximating probabilities. The SEED option allows the user to specify a random number seed; by default the program uses a value from the computer clock. The SAVE option specifies that predicted values, residuals, and model variables are to be saved to a file with the name of the file in use but with a "LAD" file extension. The SAVEd file can also be named by supplying a file name.

The QUANT = *num* | *ALL* option specifies a regression quantile, where the number specified must be greater than 0.0 and less than 1.0. Specifying QUANT = ALL yields all quantile regression estimates and when combined with a SAVE = *file name*, the parameter estimates by quantile are saved in a file with estimates (column variables) by quantiles (rows).

The HYPOTHESIS command is used to specify a reduced parameter null model against which to test the regression given by the current LAD (/QUANT = *num*) command. Note that it is not possible to test a HYPOTHESIS when all quantiles were selected with the LAD/QUANT = ALL option. The dependent variable should be the same as that on the most recent LAD command line and a reduced number of the same independent variables used in the LAD command must be given. The syntax of HYPOTHESIS is similar to LAD with NPERM and SEED options. The TEST option need not be given on the LAD command line if a HYPOTHESIS is specified. The RANKSCORE option bases hypothesis tests on a scoring function of the sign of the residuals for the reduced parameter model specified by HYPOTHESIS . Asymptotic Chi-square distributional and permutation approximations of *P*-values are both provided. The DP option provides double permutation for null models that are constrained through the origins, for either the drop in dispersion permutation test or the RANKSCORE test option. The SAVETEST = *file name* option allows the Monte Carlo resampled test statistics to be saved into a single column variable in the specified file, where the first value is always the observed test statistic value.

Terse output provided following an OUTPUT/TERSE command is the USEd file name, dependent variable name, estimated coefficients for intercept to p independent variables, observed test statistic (if HYP command), and P-value.

Ordinary Least Squares Regression (OLS)

Estimation and permutation testing alternatives for the familiar ordinary least squares regression are available with the OLS command. OLS regression estimates rates of change in conditional means. The permutation testing approaches are identical to those used for LAD regression, and are described in Kennedy and Cade (1996) and Anderson and Legendre (1999). The test statistic is similar in structure to that for LAD regression, except for OLS T_{obs} equals (sum of squared residuals for reduced parameter model − sum of squared residuals for full parameter model) / sum of squared residuals for full model. Large values of T_{obs} are evidence against the null hypothesis that the parameter(s) equal(s) zero. Our test statistic is equivalent to an F statistic without the degrees of freedom (df), which are not necessary because they are invariant under permutation. $T_{obs} \times (df$ full model/(df reduced − df full model)) = F statistic with numerator df equal to df reduced − df full model and denominator df equal to df full model. For testing all slope parameters equal zero, the dependent variable is permuted against the matrix of independent variables, and for testing partial models (subhypotheses) involving some subset of parameters, residuals from the reduced parameter, null model are permuted to the matrix of independent variables. The benefits and validity of these permutation schemes are described in Kennedy and Cade (1996) and Anderson and Legendre (1999).

We will demonstrate the OLS permutation procedure by returning to the soap scrap example previously analyzed with LAD regression (Fig. 11). Issue the following commands:

```
>USE NETER365.DAT
>OLS SOAP = CONSTANT + SPEED + LINE + LXS
```

We see in the output below that the estimate for the parameters differ slightly from those estimated with LAD regression. In particular, notice that the estimate for the interaction term of line speed and production line number is −0.176 for OLS compared to −0.210 for LAD regression.

```
                    Ordinary Least Squares Regression
Data Used
   Data file: NETER365.DAT
```

```
OLS Regression:
    SOAP = CONSTANT + SPEED + LINE +LXS
Results
   Number of observations: 27
       Dependent Variable: SOAP

   Independent variables          Regression coefficients
   CONSTANT                       7.57446455254740
   SPEED                          1.32204881288395
   LINE                           90.3908632348866
   LXS                            -0.176661427733713


      Sum of squares of the residuals: 9904.05692279731
```

We can test that the interaction coefficient for LXS equals zero by using the HYPOTHESIS command after the OLS command similar to what was done for LAD regression:

>HYPOTHESIS SOAP = CONSTANT + SPEED + LINE/ NPERM = 10000

The output indicates that there is little evidence to believe that the interaction term (LXS) differs from zero with $P = 0.1828$.

```
                  Ordinary Least Squares Regression
 Data Used
   Data file: NETER365.DAT
 HYPOTHESIS Regression:
     SOAP = CONSTANT + SPEED + LINE
 Results
    Number of observations: 27
    Dependent Variable: SOAP
    Independent variables          Regression coefficients
    CONSTANT                       27.2817939708983
    SPEED                          1.23074072291476
    LINE                           53.1291973496333

      Sum of squares of the residuals: 10713.6795013717

 Regression Evaluation:
    Ordinary Least Squares:
    SOAP = CONSTANT + SPEED + LINE +LXS
    Versus Hypothesis Model:
    SOAP = CONSTANT + SPEED + LINE

 Test Summary
      Number of permutations: 10000
         Random Number Seed: 3338701
      Observed Test Statistic: 0.817465595043978E-001
      P-value of variables in full model but not in reduced model:
                       0.182800000000000
```

The normal theory test of the same estimate yields $F_{1,23} = 1.88$ with $P = 0.184$, indicating the similarity of the permutation and normal theory probabilities for this example. Remember, that the LAD regression estimate (-0.21) and permutation test suggested that there was some evidence that the interaction term differed from zero with $P = 0.046$. As explained in Cade and Richards (1996), the one outlying value for line 2 (circled value in Fig. 11) has a studentized residual of 3.18, and though not a large outlier, has enough impact on the OLS regression estimate to reduce the magnitude of the LXS interaction term and increase the standard error of the estimate. If this outlier is deleted, the OLS estimate for the interaction term LXS becomes -0.23 with $F_{1, 23} = 3.70$ and $P = 0.068$, much more similar to the LAD estimate and test results. Minor outliers such as this one (Fig. 11) are likely to be missed or ignored in many analyses made with OLS regression and, thus, not detect nonzero effects with as much power as possible. LAD regression estimates and their permutation test are far less sensitive to the impacts of one or a few outliers (Cade and Richards 1996, Mielke and Berry 2001).

The OLS Command Syntax

The OLS command can be used to fit a variety of least squares regressions. The HYPOTHESIS command allows the specification of a reduced parameter OLS regression model to compare with the full parameter regression model specified in the main OLS command. The double permutation option for null models constrained through the origin also is provided by using a quantile = 0.50 for the binomial random sampling to randomly recenter the residuals. The regressions are run and the tests performed upon entering the OLS and HYPOTHESIS commands.

OLS *dep. var* = [CONSTANT +] *ind. var1* + *ind. var2* + ...

    [/TEST | NPERM = *num* | SEED = *num* | SAVE [= *file name*]

HYPOTHESIS *dep. var* = [CONSTANT +] *ind. var1* + *ind. var2* + ...

    [/NPERM = *num* | SEED = *num* | DP | SAVETEST [= *file name*]

Items to be supplied by the user are given in lower case in italics. Items in square brackets are optional. The vertical line (|) can be read as "or" and separates different options that can be specified. They can be specified in any order. The single variable named on the left of the equal sign is the dependent variable. The independent variables are listed and separated by plus signs to indicate the form of the regression model. If the model is to include a constant (or intercept term) the term CONSTANT must be placed right after the equal sign.

OLS options follow the slash (/) character. The TEST option causes the default test of all slope parameters equal to zero. The NPERM option allows the user to specify more or fewer permutations than the default of 5,000 used in approximating probabilities. The SEED option allows the user to specify a random number seed; by default the program uses a value from the computer clock. The SAVE option specifies that predicted values, residuals, and model variables are to be saved to a file with the name of the file in use but with a "OLS" file extension. The SAVEd file can also be named by supplying a file name.

The HYPOTHESIS command is used to specify a reduced parameter null model against which to test the regression given by the current OLS command. The dependent variable should be the same as that on the most recent OLS command line and a reduced number of the same independent variables used in the OLS command must be given. The syntax of HYPOTHESIS is similar to OLS with NPERM and SEED options. The TEST option need not be given on the OLS command line if a HYPOTHESIS is specified. The DP option provides double permutation for null models constrained through the origin. The SAVETEST option allows the Monte Carlo generated random sample of test statistics to be saved to a single column in the specified file, where the first value is the observed test statistic.

Terse output provided following an OUTPUT/TERSE command is the same as for LAD regression.

G-sample and 1-sample Goodness-of-fit Coverage Tests (COV)

The *g*-sample and 1-sample goodness-of-fit variants of the empirical coverages tests (Mielke and Yao 1988, 1990; Mielke and Berry 2001) are alternatives to the Kolmogorov-Smirnov family of tests for comparing cumulative distribution functions of continuous variates. If $x_{1/i} < ... < x_{n_i/i}$ are the n_i order statistics of the ith sample ($i = 1, ..., g$), $N =$ sum of the n_i from $i = 1$ to g, and $F_N(x) =$ (number of observed values among the N pooled values which are $\leq x$)/($N + 1$), the $n_i + 1$ coverages associated with the n_i observed values of the ith sample ($i = 1, ..., g$) are denoted by $C_{j/i} = F_N(x_{j/i}) - F_N(x_{j-1/i})$. Consider an example with 2 groups of 4 and 3 observations with order statistics $x_{1/1} < x_{2/1} < x_{3/1} < x_{1/2} < x_{4/1} < x_{2/2} < x_{3/2}$ The $N + 1 = 8$ empirical coverages are $C_{1/1} = C_{2/1} = C_{3/1} = C_{3/2} = 1/8$, $C_{4/1} = C_{2/2} = 2/8$, $C_{5/1} = 3/8$, and $C_{1/2} = 4/8$, e.g, $C_{4/1} = 5/8$ ((the number of observations $\leq x_{4/1}$)/8) - 3/8 ((the number of observations $\leq x_{3/1}$)/8) = 2/8. The coverage test statistic is a function of the absolute value of the difference between the observed coverages ($C_{j/i}$) and their expected value $(n_i + 1)^{-1}$ raised to some exponent v. The 1-sample goodness-of-fit coverage test implemented here is based on raising the absolute value of the coverages to an exponent of 1, and is equivalent to the test described by Sherman (1950). A special variant of this goodness-of-fit test for circular distributions is equivalent to Rao's (1976) spacing test for a uniform circular distribution. Probabilities under the null hypothesis are provided by a Pearson type III approximation based on the exact mean, variance, and skewness for the 1-sample goodness-of-fit test and based on Monte Carlo resampling approximations for *g*-sample tests.

We will examine a 2 group example of the coverage test by returning to the blue grouse migration example given for MRPP. Here we will examine just the migration distances (DIST) since the coverage test is limited to univariate comparisons. Quantile plots of the distances are given in Figure 15.

The commands to compare these data are:

```
>USE BGROUSE.DAT
>COVERAGE DIST * SEX/ NPERM = 10000
```

and the output is:

```
                Univariate G-Sample Empirical Coverage Test

 Data Used
          Data File: BGROUSE.DAT
    Grouping Variable: SEX
       Cover Variable: DIST

 Specification of Analysis
    Number of observations: 21
           Number of groups: 2
           Distance exponent: 1.00000000000000
     Number of iterations: 10000
        Random number seed: 3351888

 Group Summary
    Group Value                    Group Size
    3.00000000000000                   9
    4.00000000000000                  12

 Results
                Observed coverage statistic = 1.12167832167720
                Mean of coverage statistic = 0.981818181818185
     Estimated variance of coverage statistic = 0.299072126265342E-001
     Standard deviation of coverage statistic = 0.405406128993705E-003

     Observed standardized coverage statistic = 0.808734535074275
     Skewness of observerd coverage statistic = 0.455334416246216E-001
     Probability (Pearson Type III) of a larger
                or equal coverage statistic = 0.208567689286554
     Probability (Resampled) of a larger
                or equal coverage statistic = 0.193000000000000
```

The option NPERM = 10000 requested that we use 10,000 resamples for the Monte Carlo approximation of the P-values. Two probabilities are reported from this approximation, one which is the standard Monte Carlo approach of referencing the observed test statistic to those generated by the resampling, and a second which uses the resampled statistics to estimate the variance and skewness of the sampling distribution to be evaluated with the Pearson type III curve. Note their similarity here. We obtained $P = 0.193$ (Monte Carlo resampling approximation), which suggests that there is little evidence to conclude that the cumulative distributions of male and female blue grouse migration distances differ. It also is possible to get an exact enumeration of all possible permutations of the data for the coverage test statistic for small sample sizes ($n < 24$) by using the option /EXACT, which gives a $P = 0.204$ for the data in Figure 15. Interestingly, comparisons made with MRPP and $v = 1$ yield $P = 0.008$ (exact enumeration)and with MRPP and $v = 2$, $c = 2$ (permutation version of t-test) yield $P = 0.040$

(exact enumeration), both which suggest greater evidence that male and female migration distances differ.

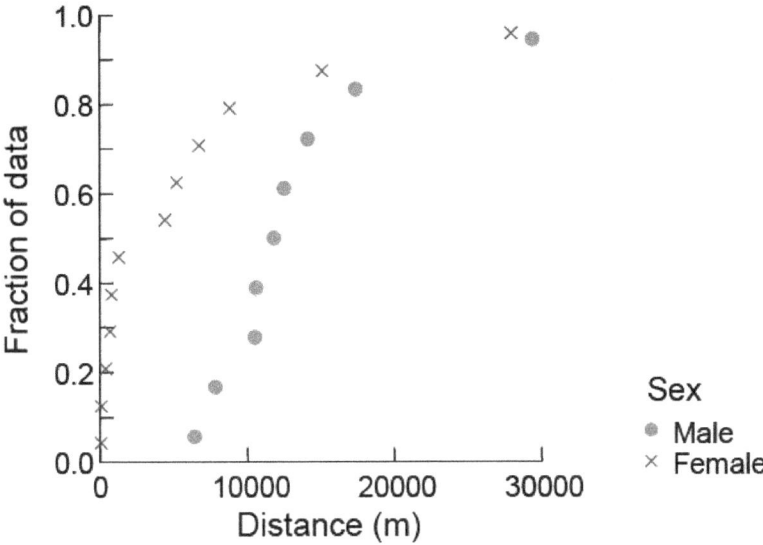

Figure 15. Quantile plots of migration distances for 9 male and 12 female blue grouse (data from Cade and Hoffman 1993).

Consider now the univariate comparisons of elevation changes (ELEV) made by male and female blue grouse in migration (Fig. 16).

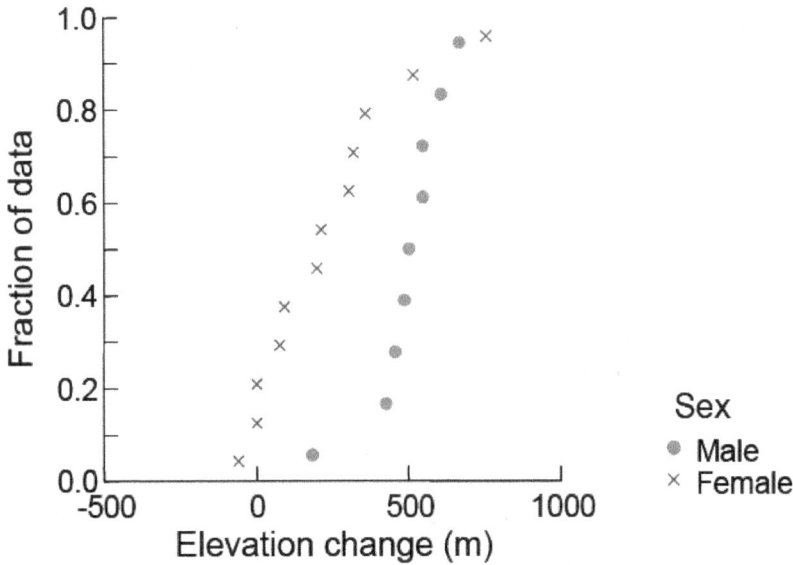

Figure 16. Quantile plots of elevation changes (m) made by 9 male and 12 female blue grouse when migrating from breeding to winter areas (data from Cade and Hoffman 1993).

We can compare these distributions with the coverage statistic by issuing the commands:

>COVERAGE ELEV * SEX/ EXACT

Here the $P = 0.019$ yields similar evidence of differences in movements as does MRPP with $v = 2$, $c = 2$ ($P = 0.010$, exact), whereas MRPP with $v = 1$ ($P = 0.004$, exact) yields slightly stronger evidence of differences.

```
            Exact Univariate G-Sample Empirical Coverage Test

  Data Used
              Data File: BGROUSE.DAT
      Grouping Variable: SEX
         Cover Variable: ELEV

  Specification of Analysis
      Number of observations: 21
            Number of groups: 2
            Distance exponent: 1.00000000000000

  Group Summary
      Group Value                 Group Size
      3.00000000000000                9
      4.00000000000000                12

  Results
                  Observed coverage statistic = 1.34755244755110
      Probability (Exact) of a larger
                  or equal coverage statistic = 0.188378185282210E-001
```

The power characteristics to detect nonzero differences between the groups clearly differ among these various test statistics. Mielke and Berry (2001) give other examples where conclusions differ greatly for coverages tests compared to MRPP comparisons. More extensive simulation research is needed to better characterize the types of distributional differences better detected by the coverage tests relative to MRPP. Simple location shifts (change in medians) appear to be detected with greater power by MRPP than by the coverage tests.

The 1-sample goodness-of-fit coverage tests are of the Kendall-Sherman type (Mielke and Berry 2001). Given an observed set of univariate data with order statistics $x_1 < x_2 < ..., < x_n$, for $i = 1$ to n these values must be transformed to the cumulative probability of the distribution function $F(x)$ specified under the null hypothesis, $U(i) = F(x_i)$ for $i = 1$ to n. These $U(i)$ are then the probability integral transformed values used in the coverage test. For example, consider the 5 values given by Bradley (1968:301-302) that were hypothesized to come from a normal distribution with mean of 3 and standard deviation of 2; -0.311, -0.078, 0.555, 1.462, and 5.711. The file BRAD302.DAT has these 5 values (X) and the 5 transformed cumulative probabilities from a normal distribution with a mean of 3 and standard deviation of 2 for these values (FX). The 1-sample goodness-of-fit test for these data are implemented by issuing the commands:

>USE BRAD302.DAT
>COVERAGE FX

The output below yield $P = 0.031$, which suggests that there is some evidence to support the belief that these 5 observations did not come from a population with a normal distribution having a mean of 3 and standard deviation of 2, similar to conclusions reached with the Kolmogorov-Smirnov goodness-of-fit test (Bradley 1968: 301-302).

```
          Kendall-Sherman Goodness of Fit Test

 Data Used
         Data File: BRAD302.DAT
    Cover Variable: FX

 Specification of Analysis
    Number of observations: 5
       Number of intervals: 5

 Results
       Observed Statistic T = 1.04951504724667
       Expected Statistic T = 0.669795953360768
    Variance of Statistic T = 0.375457304598154E-001
   Standardized statistic T = 1.95966644723580
     Skewness of statistic T = 0.246650473925624
     P-value of observed statistic,
     P(Expected >= Observed) = 0.313549281522170E-001
```

Because there are so many possible distributions that might be hypothesized and used with the 1-sample goodness-of-fit coverage test, we have not implemented any specific cumulative distribution function transformations in Blossom. We expect the user to make such transformations on the data with another statistics package prior to conducting an analysis in Blossom. There is one special cumulative distribution function transformation offered in Blossom because it is not commonly available in other statistical packages. That is we transform the data to a cumulative uniform random distribution on the unit circle to test the null hypothesis that the sample came from a population with a uniform random circular distribution. This is done with the COVERAGE test option / ARC = num, where the number provided tells the test how many units describe the circular units of measure recorded (e.g., ARC = 360 would be used for angular orientations recorded in degrees). When these transformed values are tested with the coverage test, it performs a goodness-of-fit test equivalent to Rao's (1976) spacing test for uniformity of circular distributions. We will consider the example given by Rao (1976) for the compass orientation at which 10 homing pigeons departed when released 25 km west of their loft: 20, 35, 350, 120, 85, 345, 80, 320, 280, and 85 degrees. Use the data in the file RAO.DAT.

USE RAO.DAT
COVERAGE ANGLE/ ARC = 360

The output below has $P = 0.328$ for an observed test statistic = 0.7611. This observed test statistic is related to Rao's U by $T_{obs} \times 360/2 = U$. Rao obtained $U_{10} = 137$ degrees.

```
         Kendall-Sherman Goodness of Fit Test

Data Used
        Data File: RAO.DAT
   Cover Variable: ANGLE

Specification of Analysis
   Number of observations: 10
      Number of intervals: 9
         ARC distances used
 Intervals in unit circle: 360.000000000000

Results
        Observed Statistic T = 0.761111111111111
        Expected Statistic T = 0.697356880200000
    Variance of Statistic T = 0.230444018049956E-001
   Standardized statistic T = 0.419977758737600
    Skewness of statistic T = 0.188757491268075
   P-value of observed statistic,
    P(Expected >= Observed) = 0.327828419659248
```

It is important to point out that the coverage tests assume continuous data with no tied values. The grouse examples above for the *g*-sample coverage tests had 2 tied values for both distances and elevations. This is only a minor violation of the assumption of continuity that likely has minimal impact on the analysis. At this point in time it is difficult to say what proportion of a sample comprised of tied observations constitutes a serious violation of the continuity assumption of the coverage tests. Beware of tied values.

COVERAGE Command Syntax:

The coverage command fits univariate *g*-sample empirical coverage tests if the optional grouping variable is specified or 1-sample goodness-of-fit coverage tests if no grouping variables are specified. The goodness-of-fit applications require that the user transform the observed data to be tested to the appropriate probability points from the cumulative distribution function hypothesized under the null. The exception for this rule is when the test is for a hypothesized uniform random circular distribution, where the program computes the appropriate probability integral transform based on the value selected with the option / ARC = *num*.

   COVERAGE *variable* * [*grouping variable* [*(num ...)* | *(num - num)*]] [/ EXACT |

   NPERM [*= num*]| SEED = *num* | ARC = *num* | SAVETEST [*= file name*]]

A single dependent variable is supplied by the user as indicated in italics. An optional grouping variable is specified second. The option EXACT allows for complete enumeration of the permutation sampling distribution for smaller samples (*n* < 24) and the NPERM option allows the user to select a number of resamples other than the default of 4,000 for Monte Carlo approximations of probabilities. The SEED options allows the user to specify a random number seed. The option ARC is only used for 1-sample goodness-of-fit tests where the user wants to test for uniform random circular distribution. The number specified with the ARC option is the number of units for the circular distribution (e.g., 360 degrees) to convert to a unit circle. The SAVETEST option allows the Monte Carlo generated random test statistics to be saved into a single column in the specified file, where the first value is always the observed test statistic.

Terse output provided by the COVERAGE command after an OUTPUT/TERSE command includes the USEd file name, dependent variable name, grouping variable name, number of groups, observed test statistic, and *P*-value.

110

References Cited

Aebischer, N.J., P. A. Robertson, and R.E. Kenward. 1993. Compositional analysis of habitat use from animal radio-tracking data. Ecology 74:1313-1323.

Anderson, M. J., and P. Legendre. 1999. An empirical comparison of permutation methods for tests of partial regression coefficients in a linear model. Journal Statistical Computation and Simulation 62:271-303.

Barrodale, I., and F.D.K. Roberts. 1973. An improved algorithm for discrete L1 linear approximation. SIAM Journal of Numerical Analysis 10(5):839-848.

Barrodale, I., and F.D.K. Roberts. 1974. Algorithm 478: solution of an overdetermined system of equations in the L1 norm. Communications of the Association for Computing Machinery 17:319-320.

Birkes, D., and Y. Dodge. 1993. Alternative methods of regression. John Wiley and Sons, Inc., New York. 228pp.

Berry, K.J., and P.W. Mielke. 1983. Computation of finite population parameters and approximate probability values for multi-response permutation procedures (MRPP). Communications in Statistics - Simulation and Computation12:83-107.

Berry, K.J., and P.W. Mielke. 1985. Goodman and Kruskal's tau-b statistic: A nonasymptotic test of significance. Sociological Methods and Research 13:543-550.

Berry, K.J., and P.W. Mielke 1988. A generalization of Cohen's kappa agreement measure to interval measurement and multiple raters. Educational and Psychological Measurement 48:921-933.
Berry, K.J., and P.W. Mielke. 1992. A family of multivariate measures of association for nominal independent variables. Educational and Psychological Measurement 52:41-55.

Berry, K.J., K.L. Kvamme, and P.W. Mielke. 1983. Improvements in the permutation test for the spatial analysis of the distribution of artifacts into classes. American Antiquity 48:547-553.

Biondini, M.E., C.D. Bonham, and E.F. Redente. 1985. Secondary successional patterns in a sagebrush (Artemisia tridentata) community as they relate to soil disturbance and soil biological activity. Vegetatio 60:25-36.

Biondini, M.E., P.W. Mielke, and E.F. Redente. 1988. Permutation techniques based on Euclidean analysis spaces: a new and powerful statistical method for ecological research. Coenoses 3(3):155-174.

Box, G.E.P., W.G. Hunter, and J. S. Hunter. 1978. Statistics for experimenters. John Wiley and Sons, Inc. New York. 653pp.

Bradley, J.V. 1968. Distribution-free statistical tests. Prentice-Hall, Inc. 388pp.

Cade, B.S. 1997. Comparison of tree basal area and canopy cover in habitat models: Subalpine forest. Journal of Wildlife Management 61:326-335.

Cade, B.S. 2003. Quantile regression models of animal habitat relationships. Ph.D dissertation Colorado State University, Fort Collins. 186pp.

Cade, B.S. 2005. Linear models: Permutation methods. Pages ??? *in* B. Everitt and D. Howell, eds. Encyclopedia of Statistics in the Behavioral Science. John Wiley and Sons.

Cade, B.S., and Q. Guo. 2000. Estimating effects of constraints on plant performance with regression quantiles. Oikos 91: 245-254.

Cade, B.S., and R.W. Hoffman. 1990. Winter use of Douglas-fir forests by blue grouse in Colorado. Journal of Wildlife Management 54:471-479.

Cade, B.S., and R.W. Hoffman. 1993. Differential migration of blue grouse in Colorado. Auk 110:70-77.

Cade, B.S., and B.R. Noon. 2003. A gentle introduction to quantile regression for ecologists. Frontiers in Ecology and the Environment 1: 412- 420.

Cade, B.S., and J.D. Richards. 1996. Permutation tests for least absolute deviation regression. Biometrics 52:886-902.

Cade, B.S., and J.D. Richards. In Press. A drop in dispersion permutation test for regression quantile estimates. Journal of Agricultural, Biological, and Environmental Statistics.

Cade, B.S., B.R. Noon, and C.H. Flather. 2005. Quantile regression reveals hidden bias and uncertainty in habitat models. Ecology: 86: 786-800.

Cade, B.S., J.D. Richards, and P.W. Mielke, Jr. 2005. Rank score and permutation testing alternatives for regression quantile estimates. Journal of Statistical Computation and Simulation ??:??

Cade, B.S., J.W. Terrell, and R.L. Schroeder. 1999. Estimating effects of limiting factors with regression quantiles. Ecology 80:311-323.

Dodd, C.K., and B.S. Cade. 1998. Movement patterns and the conservation of amphibians breeding in small, temporary wetlands. Conservation Biology 12:331-339.

Dodge, Y. (ed). 1987. Statistical data analysis based on the L1-norm and related methods. Elsevier Science Publishers B. V. (North-Holland). 464pp.

Dunham, J.B., B.S. Cade, and J.W. Terrell. 2002. Influences of spatial and temporal variation on fish-habitat relationships defined by regression quantiles. Transactions of the American Fisheries Society 131: 86-98.

Dyadkin, I.G., and K.G. Hamilton. 1997. A Study of 64-bit Multipliers for Lehmer Pseudorandom Number Generators. Computer Physics Communications 103: 103-130.

Freedman, D., and D. Lane. 1983. A nonstochastic interpretation of reported significance levels. Journal Business and Economic Statistics 1:292-298.

Edgington, E.S. 1987. Randomization tests. Marcel Dekker, Inc., New York. 341pp.

Fawcett, R.F. 1990. The alignment method for displaying and analyzing treatments in blocking designs. American Statistician 44:204-209.

Gentle, J.E. 1977. Least absolute values estimation: an introduction. Communications in Statistics -Simulation and Computation B6(4):313-328. (See also the rest of this volume).

Good, P. 2000. Permutation tests: A practical guide to resampling methods for testing hypotheses. 2nd edition. Springer. 270pp.

Haire, S. L., C.E. Bock, B.S. Cade, and B.C. Bennett. 2000. The role of landscape and habitat characteristics in limiting abundance of grassland nesting songbirds in an urban open space. Landscape and Urban Planning 48(1-2):65-82.

Hodges, J.L., and E.L. Lehmann. 1962. Rank methods for combination of independent experiments in analysis of variance. Annals of Mathematical Statistics 33:482-497.

Iyer, H.K., K.J. Berry, and P.W. Mielke. 1983 Computation of finite population parameters and approximate probability values for multi-response randomized block permutation procedures (MRBP). Communications in Statistics - Simulation and Computation 12(4):479-499.

Kennedy, P.E., and B.S. Cade. 1996. Randomization tests for multiple regression. Communications in Statistics - Simulation and Computation 25:923-936.

Koenker, R. 1994. Confidence intervals for regression quantiles. *Pages* 349-359 *in* P. Mandl and M. Hušková, editors. Asymptotic statistics: Proceedings of the 5th Prague Symposium. Physica-Verlag.

Koenker, R., and G. Bassett. 1978. Regression quantiles. Econometrica 46:33-50.

Koenker, R., and G. Bassett. 1982. Robust tests for heteroscedasticity based on regression quantiles. Econometrica 50:43-61.

Koenker, R., and J.A.F. Machado. 1999. Goodness of fit and related inference processes for quantile regression. Journal of the American Statistical Association 94:1296-1310.

Koenker, R., and S. Portnoy. 1996. Quantile regression. University of Illinois at Urbana-Champaign, College of Commerce and Business Administration, Office of Research Working Paper 97-0100. 77pp.

Manly, B.F.J. 1991. Randomization and Monte Carlo Methods in Biology. Chapman and Hall, New York. 281pp.

Matsumoto, M., and T. Nishimura. 1998. Mersenne Twister: A 623-Dimensionally equidistributed uniform pseudo-random number generator. ACM Transactions on Modeling and Computer Simulation 8:3-30.

Mielke, P.W. 1984. Meteorological applications of permutation techniques based on distance functions. Pages 813-830 in P.R. Krishnaiah and P.K. Sen, eds. Handbook of statistics, Vol. 4. North-Holland, Amsterdam.

Mielke, P.W. 1985. Geometric concerns pertaining to applications of statistical tests in the atmospheric sciences. Journal of Atmospheric Sciences 42:1209-1212.

Mielke, P.W. 1986. Non-metric statistical analyses: some metric alternatives. Journal of Statistical Planning and Inference 13:377-387.

Mielke, P.W., Jr. 1991. The application of multivariate permutation methods based on distance functions in the earth sciences. Earth-Science Reviews 31:55-71.

Mielke, P.W., and K.J. Berry. 1982. An extended class of permutation techniques for matched pairs. Communications in Statistics - Theory and Methods A11:1197-1207.

Mielke, P.W., and K.J. Berry. 1983. Asymptotic clarifications, generalizations, and concerns regarding an extended class of matched pairs tests based on powers of ranks. Psychometrika 48:483-485.

Mielke, P.W., and K.J. Berry. 1994. Permutation tests for common locations among samples with unequal variances. Journal Educational and Behavioral Statistics 19:217-236.

Mielke, P.W., Jr., and K.J. Berry. 1999. Multivariate tests for correlated data in completely randomized designs. Journal Educational and Behavioral Statistics 24:109-131.

Mielke, P.W., Jr., and K.J. Berry. 2001. Permutation methods: A distance function approach. Springer-Verlag. 352pp.

Mielke, P.W., and H.K. Iyer. 1982. Permutation techniques for analyzing multi-response data from randomized block experiments. Communications in Statistics - Theory and Methods 11:1427- 1437.

Mielke, P.W., and Y.C. Yao. 1988. A class of multiple sample tests based on empirical coverages. Annals of the Institute of Statistical Mathematics 40:165-178.

Mielke, P.W., and Y.C. Yao. 1990. On g-sample empirical coverage tests: Exact and simulated null distributions of test statistics with small and moderate sample sizes. Journal Statistical Computation and Simulation 35:31-39.

Mielke, P.W., K.J. Berry, and G.W. Brier. 1981a. Application of multi-response permutation procedures for examining seasonal changes in monthly mean sea-level pressure patterns. Monthly Weather Review 109:120-126.

Mielke, P.W., K.J. Berry, and E.S. Johnson. 1976. Multi-response permutation procedures for a priori classifications. Communications in Statistics - Theory and Methods A 5:1409-1424.

Mielke, P.W., K.J. Berry, and J.G. Medina. 1982. Climax I and II: Distortion resistant residual analyses. Journal Applied Meteorology 21-788-792.

Mielke, P.W., K.J. Berry, P.J. Brockwell, and J.S. Williams. 1981b. A class of nonparametric tests based on multiresponse permutation procedures. Biometrica 68:720-724.

Mielke, H.W., J.C. Anderson, K.J. Berry, P.W. Mielke, R.L. Chaney, and M. Leech. 1983. Lead concentrations in inner-city soils as a factor in the child lead problem. American Journal of Public Health 73:1366-1369.

Narula, S.C., and J.F. Wellington. 1982. The minimum sum of absolute errors regression: State of the art survey. International Statistical Review 50:317-326.

Pielou, E.C. 1984. The interpretation of ecological data. John Wiley and Sons, New York. 263pp.

Rao, J.S. 1976. Some tests based on arc-lengths for the circle. Sankhya, Series B 38:329-338.

Reich, R.M., P.W. Mielke Jr., and F.G. Hawksworth. 1991. Spatial analysis of Ponderosa pine trees infected with dwarf mistletoe. Canadian Journal Forest Research 21:1808-1815.

Schroeder, R.L., and L.D. Vangilder. 1997. Tests of wildlife habitat models to evaluate oak mast production. Wildlife Society Bulletin 25:639-646.

Seneta, E. 1983. The weighted median and multiple regression. Australian Journal of Statistics 25:370-377.

Sherman, B. 1950. A random variable related to the spacing of sample values. Annals of Mathematical Statistics 21:339-361.

Slauson, W.L. 1988. Graphical and statistical procedures for comparing habitat suitability data. U.S. Fish and Wildlife Service, Biological Report 89(6).

Solow, A.R. 1989. A randomization test for independence of animal locations. Ecology 70:1546-1549.

Terrell, J.W., B.S. Cade, J.Carpenter, and J.M. Thompson. 1996. Modeling stream fish habitat limitations from wedged-shaped patterns of variation in standing stock. Transactions of the American Fisheries Society 125:104-117.

Tucker, D.F., P.W. Mielke, Jr., and E.R. Reiter. 1989. The verification of numerical models with multivariate randomized block permutation procedures. Meteorology and Atmospheric Physics 40:181-188.

Van Valen, L. 1978. The statistics of variation. Evolutionary Theory 4:33-43.

Whaley, F.S. 1983 The equivalence of three independently derived permutation procedures for testing the homogeneity of multidimensional samples. Biometrics 39:741-745.

Wong, R.K.W., N. Chidambaram, and P.W. Mielke. 1983. Application of multi-response permutation procedures and median regression for covariate analyses of possible weather modification effects on hail responses. Atmosphere-Ocean 21:1-13.

Zimmerman, G.M., H. Goetz, and P.W. Mielke. 1985. Use of an improved statistical method for group comparisons to study effects of prairie fire. Ecology 66:606-611.

Appendix A · Common Statistical Tests Embraced by the MRPP Command

Multiresponse permutation procedures (MRPP) can duplicate many common statistical tests (the parametric tests listed below are all permutation versions).

Two-sample t-test
One-way analysis of variance
Multivariate analysis of variance
Hotelling's T^2
Median test (2 and k-sample)
Wilcoxon-Mann-Whitney test
Kruskal-Wallis test
Goodman and Kruskal contingency table tests of association
      (tau-a, tau-b)
Generalized runs tests (including Wald-Walfowitz runs test)
Durbin-Watson for univariate first-order autoregression
Schoener's t^2/r^2 for bivariate first-order autoregression

The multiresponse permutation procedures for randomized block data (MRBP) and permutation tests for matched pairs (PTMP) can duplicate the following tests (the parametric tests listed below are all permutation versions).

Matched pairs and 1-sample t-test
Analysis of variance for complete randomized blocks
Sign test
Wilcoxon signed rank test
Pearson correlation coefficient
Spearman rank correlation
Kendall tau (correlation)
Friedman's test for randomized blocks
Spearman's footrule and multi-block extension
Cochran's Q and McNemar's tests
Cohen's kappa

Other less familiar tests are also known to be special cases of MRPP and MRBP. Be aware, too, that many of the above tests are strictly univariate or bivariate, but MRPP and MRBP often generalize to the multivariate case as well. Further, most of the above listed tests use the square of Euclidean distance in the definition of the test statistic, whereas MRPP and MRBP have the option of choosing a distance measure commensurate with the data space. The generalized distance function in MRPP yields alternative, often more powerful versions of these tests.

Appendix B - Blossom Statistics Program Installation, Configuration, Requirements

The README files included with Blossom has the latest information for configuring and installing Blossom as well as descriptions of recent program updates and improvements. The Blossom website at http://www.fort.usgs.gov/products/software/blossom/blossom.asp has the latest information and releases of the program.

Blossom Installation Program

The Blossom installation program gives basic instructions that should be easy to follow. The default installation folder will be in the "Program Files" folder of the first local drive (hard disk directory). An alternate installation path can be provided. The installation program creates a "Blossom" folder and installs files and folders needed to run Blossom. If the folder already exists, you can overwrite it. This is an easy way to update Blossom software. A dialog box appears asking to add programs to the Windows Start menu and to add environment variables. Blossom will need the environment variables for proper operation. Once done, the computer may need to be re-booted to run Blossom.

Some versions of Windows use the Windows registry. Appropriate paths for an environment variable called "BLOSSDIR" are added to the registry as well as an addition in the "PATH" variable to the Blossom installation folder. Other versions of Windows find environment variables set in the "AUTOEXEC.BAT" file, and appropriate lines will be added to that file so that Blossom will run.

Blossom Configuration

Two Blossom programs

Blossom is installed with two executable programs, a Console version (CONBLOS.EXE) and a Windows version (BLOSSOM.EXE). The Console version runs best from a Command Prompt window configured as described below. It appears as a "glass teletype" and opens no true windows. All interaction with the Console version is from the command line. The Windows version of Blossom has some Windows functionality, but it also can interact with the user from a command line data entry window. Aside from the user interface, both programs use identical code so results should be identical.

The Console version of Blossom does not have additional Windows overhead and so runs slightly faster. It does use Windows for virtual memory management, so that data file size limitations depend on how much Windows Virtual Memory is available. The Console version is more amenable to processing very large Blossom "submit" files (ASCII text files of Blossom commands). This makes it particularly well suited for batch processing of many analyses of many data sets (as in a simulation study).

The Windows version of Blossom runs nearly as quickly as the Console version, yet has additional Windows functionality for accessing files, obtaining Windows help for Blossom, and selecting, clipping and printing output. This version is best suited for interactive sessions where most of the time is spent considering what needs to be done next and inspecting results. Access to the file system is simplified through interaction with file access dialog boxes. The Windows version also processes Blossom (submit) command files.

Windows version configuration

The Windows version of Blossom is installed to run from the Windows Start Programs menu. This may be copied and pasted as a shortcut onto the Windows Desktop. No further configuration is necessary. When first installed, Windows Blossom program is run from the installation BLOSSOM\BIN folder and accesses files from the installation BLOSSOM\SAMPLES folder. Sample files in that folder can be accessed with the USE and SUBMIT commands to follow the examples in this manual (see SUBMIT a Command File and USE a Data File in the General Program Functions section, above). Once this version is started, you can change folders to use data files or submit command files. Accessing a data file or submit command file (without specifying the name) initiates a file access dialog box that allows navigation within the file system. Blossom keeps track of the current folder and restarts in the last working folder of the previous session.

When Blossom is run, a small entry field opens at the bottom of the Blossom window. This is the "Blossom Command>" or Blossom command prompt window. Commands to Blossom are entered here. The last 100 commands of a session are kept by Blossom. Pressing the F3 function key and highlighting the command to access recalls previous commands. A command can be edited in the Blossom command prompt window using normal Windows editing functions. When editing is done, press the ENTER key to send the command to Blossom for processing.

The Windows version of Blossom can be started with optional operating system command line arguments. The properties of a shortcut on the Windows Desktop (right click on it and select Properties), can be edited to provide a data file for Blossom to use, followed by the Blossom command to perform a statistical analysis. Blossom starts up, uses the data file, and performs the analysis. Alternatively, a SUBMIT command can be added (see SUBMIT a Command File in this document). To the invocation of the Blossom program, append the arguments "SUBMIT" and then the name of the file to submit. Blossom starts up and processes commands in the command file. Another, possibly more useful way to do this is to invoke the Windows version of Blossom with command line arguments from a Windows Command Prompt window (sometimes called a "DOS Box" or DOS Window). Operating system command line arguments is discussed in the following section. The BLOSSOM.EXE program can be invoked in the same way as CONBLOS.EXE from a "DOS prompt".

Console version configuration

After installation, the Console version of Blossom can be run from the Windows Start menu. All access to Blossom is through the Blossom command line, entered at the Blossom command prompt, which is a "greater than sign" or "right angle bracket" (">").

As a Console program, this version of Blossom has access to files in the local folder where it is invoked, the "Start In" folder in Windows terminology. The Console program resides in the installation BLOSSOM\BIN folder and no data are there, so not much can be done. If Properties of the Console program shortcut are accessed (right-click on the program title on the Start menu and select "Properties"), the "Start In" folder can be modified to where data to be analyzed can be found. It is instructive to first set this to the installation BLOSSOM\SAMPLES folder. When Console Blossom is run; then access is to data and Blossom SUBMIT) command files that are discussed in this manual. Change the "Start In" folder of the shortcut properties to access other folders where data are to be analyzed.

The Console Blossom shortcut from Windows Start menu can be copied to the Windows Desktop. Changing the "Start In" folder of a shortcut makes a separate shortcut to each data folder. Multiple shortcuts can be consolidated within a folder on Windows Desktop.

An alternate and flexible way to run the Console version of Blossom is to set up a Command Prompt Window (sometimes called a "DOS Box" or DOS Window). The usual Windows icon for this is a stylized "MSDOS". Depending on the version of Windows, this invokes a session of COMMAND.COM or CMD.EXE. Access to the file system and other DOS commands is from the Command Prompt Window as well as CONBLOS.EXE (the Console Blossom).

Properties of the Command Prompt Window can be modified. Depending on Windows version, the number of lines displayed can be altered (something like "Properties, Layout, Window size, height" may be adjusted). The screen can be adjusted to a scrollable window (with any number of lines in the virtual screen size). This adjustment can be made by accessing something like "Properties, Layout, Screen Buffer size, height". As in all Command Prompt Windows, there are methods available to copy and paste using the Windows Clipboard. This is useful for pasting repetitive or complicated commands or lists of variable names. Windows documentation should be consulted for specific information about changing the properties of a Command Prompt Window.

The Console version of Blossom can be run with operating system command line arguments, i.e., additional information can be sent to the Blossom program to begin processing. For example, from the Windows Command Prompt Window's prompt (the "DOS prompt"), the command

C:\PROGRA~1\Blossom\samples>CONBLOS MRBP.DAT

invokes Console Blossom and accesses MRBP.DAT data file, just as the USE command would. Moreover, a statistical command can be added to the command line and the program will access the data file and execute the statistical command, so:

```
C:\PROGRA~1\Blossom\samples>CONBLOS MRBP.DAT MRPP SPP1 SPP2
    SPP3*TRTMT*BLOCK
```

runs Console Blossom, USEs data file MRBP.DAT, and performs a multiresponse randomized block procedure.

Blossom programs can process Blossom (SUBMIT) command files invoked from the "DOS prompt". The command:

```
C:\PROGRA~1\Blossom\samples>CONBLOS SUBMIT SUBWAY.SUB
```

runs Console Blossom and SUBMITs the Blossom command file SUBWAY.SUB.

Details about USE and SUBMIT as well as operating system command line arguments are given in the **General Program Functions** section of this manual (see **SUBMIT a Command File**, **USE a Data File**, and **Advanced SUBMIT Operations with Program Arguments and DOS Batch Files sections**).

Blossom Requirements and Program Limits

Blossom is compiled to run in a 32-bit Windows environment. Both the Console and Windows Blossom should run under Windows versions XP and 2000. The program is not guaranteed to run under Windows 95, 98, ME, or NT as these operating systems are no longer supported in our environment. Development and testing was done originally under Windows 98 and NT but for the last few years we have used only Windows 2000 and XP. Both Blossom versions are Window programs, including the Console version (despite its user interface), and depend on Windows for various functions, including virtual memory management. No Real Mode virtual memory manager is bound into Blossom, so the programs are not expected to operate in true DOS mode, although the Console readily runs in a "DOS Window" (also known as "DOS Box" or Command Prompt Window) opened from Windows.

Blossom installation takes approximately 6 Megabytes of disk space, including programs, documentation, help and support files, and sample datasets.

During operation, Blossom creates a copy of any data file USEd in a unique Blossom file format. This file (TEMP-DAT.$$$) contains values of numeric data from the USEd data file in binary format. The size (in bytes) of this file can be computed approximately by multiplying 8 byte number of observations by the number of variables. Space should be available on disk for this file to exist (plus space for output files and the Blossom history file). For example, a file with

3450 observations and 14 variables takes approximately 386400 bytes (actually it takes 386512 bytes due to some system-dependent formatting overhead). If there are missing values in data used, a duplicate temporary file may be created with appropriate records dropped for the analysis, so the space requirement is doubled. With multiresponse randomized block procedures (MRBP), a file may be created as a subset of the original file with the grouping a blocking structures of the variables used. The maximum temporary MRBP space requirement is the size for the non-missing data.

Blossom allocates virtual memory space dynamically at runtime. This means that the amount of memory required by Blossom depends on the program analyses being run and the size of the data and associated internal storage required for the analysis. If this exceeds the physical Random Access Memory (RAM) available, Blossom uses Windows virtual memory management. This runtime memory required cannot exceed the paging file size (swap space) available to Windows. Documentation to Windows virtual memory management should be consulted.

In the Windows version of Blossom, part of the (virtual) memory used includes space used by the output screen where the commands are echoed and results of the statistical analyses are written by default. This can be minimized by using the ECHO OUTPUT=OFF command (which stops statistical results being written to screen) and can be temporarily minimized by issuing the CLS (clear output screen) command.

Internally, Blossom has some limits on the amount of data it can support. The total amount of memory the program and dynamically allocated array space can occupy is about 2 gigabytes. Here are some limits within Blossom:

- Number of elements in a Blossom command: 1024 (command plus all variables and options and delimiters).
- Maximum single command element length in ASCII representation: 25 bytes
- Maximum size of Blossom command: 8192 bytes
- Number of variables: 1024
- Number of observations: about 2 billion (depends on number of variables, total memory limited to about 2GB)
- Number of quantiles in a Median and Quantile (MEDQ) analysis: 250,000,000
- Number of observations in a MEDQ: 250,000,000
- Number of groups in one MEDQ: 250,000,000
- Maximum group size in MEDQ: 250,000,000
- Maximum number of variables in a MEDQ analysis: 255
- Maximum number of blocks in an Exact Multiresponse Randomize Block Procedure: 9
- Data file size: Limit of Windows Virtual Memory (depends on associated memory required for an analysis)
- Filename size: 25 bytes
- Variable name size: 25 bytes (12 characters for SYSTAT files)
- Title length (optional, for labeling statistical results output): 80 bytes
- Maximum significant digits of double precision numbers: about 15

• Missing value internal representation: 0.10 X 10-37 (no datum within a data file should have this value)

Note that there may be other, smaller limits depending on the combination of all factors considered.

Some specific Blossom statistical analyses require minimum numbers of elements:

• Minimum number of observations for Multi-Response Permutation Procedure (MRPP): 6
• Minimum number of groups for MRPP: 2
• Minimum group size for MRPP: 2
• Minimum number of observations for Exact MRPP: 3
• Minimum number of groups for an Exact MRPP: 2
• Minimum group size for Exact MRPP: 2
• Minimum number of groups for Multiresponse Randomized Block Procedure (MRBP): 2
• Minimum number of groups for an MRBP: 2
• Minimum number of blocks for Exact MRBP: 2
• Minimum number of cases for Permutation Test for Matched Pairs (PTMP): 3
• Minimum number of observations for Multi-Response Sequence Procedure (MRSP): 6
• Minimum number of observations for Exact MRSP: 2

Appendix C - Blossom Development and Testing

We consider Blossom to be a platform for supporting research into various statistical concepts. It provides for data access and display of results for statistical functions we are investigating. We don't consider it to be a general purpose statistical package, nor do we make any assertions about its ease of use or efficacy for statistical computing.

Blossom is written in Fortran. The current version is in Fortran 95, using the Lahey/Fujitsu LF95 Version 7.1.2 compiler (Win32 Professional Language System), from Lahey Computer Systems, Inc.

Blossom access to Microsoft Windows is made possible with the Winteracter Fortran 9x GUI Toolset from Interactive Software Services, Ltd.

Blossom can use either of two pseudo random number generators (PRNG). By default the PRNG used in Blossom is provided by the Lahey/Fujitsu Fortran compiler. This pseudorandom number generator technology is based on Lehmer's pure multiplicative congruential algorithm (Dyadkin and Hamiltion 1997). Alternatively, upon user specification, Blossom uses the Mersenne Twister (MT) PRNG (Matsumoto and Nishimura 1998). Matsumoto and Nishimura describe Mersenne Twister as a twisted generalized feedback shift register sequence (TGFSR) algorithm.

Hundreds of hours of testing have gone into this (and previous) version of Blossom since development was initiated in 1989. We have done everything possible to ensure that our modifications of the principal computing routines for the MRPP family of statistics obtained from Paul. W. Mielke, Jr., yield numerical results identical to the original routines. We made comparisons with published and other known results, and had Dr. Mielke compare some of his original analyses with our program. Similar comparisons of numerical output were made for the regression quantile and rank score tests made with programs provided by Roger Koenker. No doubt, some errors remain undetected and we urge you to report any obvious or suspicious errors to us. The README file provides a list of pertinent information to provide us in an error report. We recommend running analyses on the datasets provided with the Blossom software to see if your computer duplicates the output in the User Manual and output files.

The Blossom software we provide has been scanned for computer viruses using the latest versions of Symantec anti-virus scanning programs. As far as we are able to determine, Blossom software as distributed from us is virus-free and there are no malware components. Blossom is not, we are sure, free of bugs. The authors welcome reports of errors.

Appendix D · Acknowledgements

We would like to acknowledge the continual interest and support provided by Drs. Mielke and Koenker, and the testing work done on previous versions by Dr. David R. Smith. We thank Dr. William L. Slauson for initiating the first version of the Blossom User's Manual that this one replaces.

www.ingramcontent.com/pod-product-compliance
Lightning Source LLC
Chambersburg PA
CBHW082031190526
45166CB00017B/2477